CLASSICAL
GALOIS
THEORY

WITH EXAMPLES

63-57

CLASSICAL
GALOIS
THEORY

WITH EXAMPLES

LISL GAAL

UNIVERSITY OF MINNESOTA

MARKHAM
PUBLISHING COMPANY

CHICAGO

MARKHAM MATHEMATICS SERIES
William J. LeVeque, Editor

Anderson, *Graph Theory and Finite Combinatorics*
Cantrell and Edwards, eds., *Topology of Manifolds*
Gaal, *Classical Galois Theory with Examples*
Gioia, *The Theory of Numbers: An Introduction*
Greenspan, *Introduction to Numerical Analysis and Applications*
Hu, *Calculus*
Hu, *Cohomology Theory*
Hu, *Elementary Functions and Coordinate Geometry*
Hu, *Linear Algebra with Differential Equations*
Knopp, *Theory of Area*
Knopp, *Modular Functions in Analytic Number Theory*
Peterson, *Foundations of Algebra and Number Theory*
Stark, *An Introduction to Number Theory*

LECTURES IN ADVANCED MATHEMATICS

Davenport, *1. Multiplicative Number Theory*
Storer, *2. Cyclotomy and Difference Sets*
Engeler, *3. Formal Languages: Automata and Structures*
Garsia, *4. Topics in Almost Everywhere Convergence*

Copyright © 1971 by Markham Publishing Company
All rights reserved
Printed in the United States of America
Library of Congress Catalog Card Number: 76-91019
Standard Book Number: 8410-1907-X

TO MY PARENTS

DR. JOSEF AND HERTHA H. NOVAK

PREFACE

Galois theory is one of the most beautiful subjects in mathematics, but it is hard to appreciate this fact fully without seeing specific examples. Numerous examples are therefore included throughout the text, in the hope that they will lead to a deeper understanding and genuine appreciation of the more abstract and advanced literature on Galois theory.

The book is intended for beginning graduate students who already have some background in abstract algebra, including some elementary theory of groups, rings, and fields.

Chapter I consists of results that should be used as background references of definitions, theorems, and examples. The study of Galois theory proper begins in Chapter II, which introduces automorphisms of fields and related topics. In Chapter III we define normal extensions, prove the fundamental theorem, and give examples that illustrate what happens here. Chapter IV contains applications. We prove that an equation is solvable if and only if its group is solvable, taking the meaning of "solvable" in its strict sense—that a solution can actually be produced and written down. Since the actual solution of a solvable polynomial requires that we first determine the group, we also include a procedure for doing this. The procedures that accomplish this are complicated, as one would expect of a process that must cover the solution of all possible solvable equations, but nonetheless it is an algorithm, as is the procedure for finding the group of a polynomial. Among the other

PREFACE

applications we include ruler-and-compass constructions, a proof that in general the field $Q(\sqrt[n]{p_1}, \ldots, \sqrt[n]{p_k})$ is of degree n^k over the rational field Q ($p_1, \ldots, p_k$ are primes), and a proof that every Abelian extension of Q is a subfield of some cyclotomic extension of Q.

The exposition and proofs are intended to present Galois theory in as simple a manner as possible, sometimes at the expense of brevity. Occasionally this also involves a certain amount of repetition. The book is for students and intends to make them take an active part in mathematics rather than merely read, nod their heads at appropriate places, skip the exercises, and continue on to the next section. This is why there are often blanks in the examples and even in some proofs. These are to be filled in by the reader and will check how well he understands the material. Many, but not all, can be filled in on inspection. To encourage further reading there are many references, especially to books that are also on the beginning graduate level, but there is no intention of including the entire literature of Galois theory. The now classic book by Artin (*Galois Theory*, Notre Dame University, 1942) would be a valuable companion book, especially for its beautifully condensed presentation of Galois theory. References are enclosed in brackets and refer to the bibliography at the end.

The book is based in part on a seminar given in the summer of 1967 at the University of Colorado and on a course in algebra at the University of Minnesota in 1965–1966 and 1969–1970. I am very grateful to my students, who read various parts of the manuscript and suggested changes and corrections, in particular Mr. W. McMullen, Mr. D. McPhaden, Mr. F. Minbashian, and Mrs. M. Skoog, who worked out the example presented in Sections 4.8 and 4.9, and also to my colleague Professor I. J. Richards of the University of Minnesota. For much needed encouragement I also thank Mrs. D. Aeppli, and especially my husband.

University of Minnesota   LISL GAAL
Minneapolis, Minnesota

CONTENTS

CONTENTS

CHAPTER I

PREREQUISITES

This chapter is intended to be a summary of some of the algebraic background we shall need later. Examples are included, but proofs are generally omitted. They may be found in any standard reference, and a number of these are cited. It is assumed that the reader is already familiar with this material and will use Chapter I as a reference to be consulted when, or rather if, necessary in reading the text from Chapter II on, and also to check on the notations and definitions that are being used, because a number of concepts can be defined in several different ways (for example, solvable groups or ideals, which might be left-, right-, or two-sided). Although Chapter I is not an exposition of the background suitable for those who have never seen the material, examples are included to help with the reviewing and perhaps gain a somewhat deeper insight into these concepts than at the first meeting.

1.1 GROUP THEORY

DEFINITION.   An algebraic system $\langle G, *, e \rangle$ is called a *group* if G is a set of elements on which a binary operation $*$ is defined and e is an element of G so that the following conditions are satisfied for all $x, y, z \in G$:

(1) $x * y \in G$ (closure of G under $*$).

1

1.1 GROUP THEORY

(2) $x * (y * z) = (x * y) * z$ (associative law).

(3) $x * e = e * x = x$ (identity element).

(4) For every $x \in G$ there is a $y \in G$ such that $x * y = y * x = e$ (existence of inverses).

The element e is called the *identity* of G, and it is easy to show that G contains no other element which satisfies (3) for every x and that the inverse of every element is uniquely determined.

DEFINITION. A group is called *Abelian* if in addition to (1) through (4) we also have

(5) $x * y = y * x$, for every $x, y \in G$ (commutative law).

It usually causes no confusion to denote the group by G rather than $\langle G, *, e \rangle$. In general, the letters G and H are used to denote groups.

NOTATION. If S is a set, then $|S|$ denotes the number of elements in S. (If S is infinite, then $|S|$ denotes the cardinal of S.)

DEFINITION. If G is a group, then $|G|$ is called the *order* of G.

DEFINITION. The set H is a *subgroup* of G if (1) $H \subseteq G$ and (2) H is a group. We shall write $H < G$ or $G > H$. A subgroup H is called a *proper subgroup* of G if $H < G$ and $\{e\} \neq H \neq G$. (Sometimes we write $H \leq G$ or $H \not\leq G$ to emphasize that H is, or is not, allowed to equal G.)

The group consisting of the identity only is $\{e\}$. Later, when we deal with permutations, this group is denoted by (1).

Subgroups of G form a lattice under inclusion: For example, we can list all the subgroups of Z_6^+ (the group of integers mod 6 under $+$) and arrange them as follows:

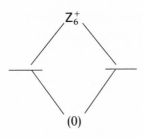

In Table 1.1 you can check which of the group axioms hold for the given sets and operations, and also list some subgroups, if the system turns out to be a group. Here Z is the set of positive and negative integers, Z_n the integers mod n, Z_n^+ the group $\langle Z_n, +, 0 \rangle$, and Q the field of rationals.

TABLE 1.1   Examples for Group Axioms

Set G	Operation *	Closed	Associative	Identity element	Inverse	Commutative	Is it a group?	List some proper subgroups; remarks
Z	+							
Z	−	√	No	No	−	No	No	
Z	.							
Z	÷							
Z_n	+							
Z_2	−							
Z_n	.							
$Z_n - \{[0]\}$	.							
Q	+							

1.1 GROUP THEORY

TABLE 1.1 Examples for Group Axioms (contd.)

Set G	Operation $*$	Closed	Associative	Identity element	Inverse	Commutative	Is it a group?	List some proper subgroups; remarks
Q	.							
$Q - \{0\}$	.							
$\{0, 1\}$	Exponen-tiation a^b (see →)							Table for a^b: (see table)
$A(S)$ (all $1 \sim 1$ mappings of the given set S onto itself)	Composition of mappings							
$\left\{ r, \dfrac{1}{r}, 1 - r, \dfrac{1}{1 - r}, \dfrac{r}{r - 1}, \dfrac{r - 1}{r} \right\}$	Substitution (functional composition: e.g., $\left(\dfrac{1}{r}\right) * (1 - r) = \left(1 - \dfrac{1}{r}\right).$)							

Table for a^b:

$b \backslash a$	0	1
0	1	1
1	0	1

DEFINITION. If $H < G$, then the set $Hx = \{hx | h \in H\}$ is called a *left coset* of H in G. Similarly, xH is a *right coset*.

DEFINITION. The *index of H in G* (ind$_G$ H) is the number of distinct (left) cosets of the subgroup H in G.

DEFINITION. If $H < G$, then H is a *normal or invariant subgroup of G* if $xH = Hx$ for every x. We shall write this $H \lhd G$ or $G \rhd H$.

DEFINITION. If $H \lhd G$, then G/H is the set of distinct cosets xH, where x, y, ... range over G and we define an operation $*$ as follows: $(xH)*(yH) = (xy)H$. With this operation, G/H is a group called the *factor* or *quotient group* of G by H.

EXERCISE. Show that in an Abelian group every subgroup is normal. Then find the factor groups of Z_6^+ by its subgroups.

DEFINITION. The mapping $\varphi: G_1 \to G_2$ is a *homomorphism* from G_1 to G_2 if for every x, $y \in G$ we have $\varphi(xy) = (\varphi x) \cdot (\varphi y)$.

DEFINITION. The groups G and H are *isomorphic* (written $G \cong H$) if there is a one-to-one mapping φ of G onto H (bijection) such that $\varphi(xy) = (\varphi x) \cdot (\varphi y)$ for all $x, y \in G$.

EXERCISES. (1) Check that isomorphism is an equivalence relation.

(2) Show that

(a) $\langle Z_6, +, 0 \rangle \cong \langle Z_7 - \{0\}, \cdot, 1 \rangle$.

(b) $\langle Z_4, +, 0 \rangle \ncong \langle Z_8 - \{0, 2, 4, 6\}, \cdot, 1 \rangle \rangle$.

In (a) the isomorphism is established by the correspondence $f: Z_6 \to (Z_7 - \{0\})$ in the table

Z_6	0	1	2	3	4	5
$Z_7 - \{0\}$						

(3) Suppose $H \nleq G$. Then it $\begin{cases} \text{is} \\ \text{is not} \end{cases}$ (cross out one) possible that $H \cong G$, for example, take $G = $ _____, $H = $ _____.

1.1 GROUP THEORY

DEFINITION. If $H \lhd G$, then H is called a *maximal normal* subgroup of G (written $H \underset{\text{max}}{\lhd} G$), if (1) $\{e\} \neq H \neq G$ and (2) there is no normal subgroup K of G such that $H \lneqq K \lneqq G$.

DEFINITION. Given a group G, a sequence of groups $G_0, \ldots, G_r$ is called a *composition series* for G if

$$G = G_0 \underset{\text{max}}{\rhd} G_1 \underset{\text{max}}{\rhd} G_2 \cdots G_{r-1} \underset{\text{max}}{\rhd} G_r = \{e\},$$

where G_{r-1} has no proper normal subgroup. The groups G_i/G_{i+1} are called *composition factors* of G, their number r is the *length* of the series, and the integers $|G_k/G_{k-1}|$ are the *composition orders*.

DEFINITION. Two composition series for G are called *isomorphic* if they have the same length and the composition factors are pairwise isomorphic (but they do not necessarily appear in the same order).

For example, there are two composition series for Z_6^+. They are

$$Z_6^+ \underset{\text{max}}{\rhd} \underline{\hspace{3cm}} \underset{\text{max}}{\rhd} (0) \quad \text{and} \quad Z_6^+ \underset{\text{max}}{\rhd} \underline{\hspace{3cm}} \underset{\text{max}}{\rhd} \{0\}.$$

The factors are isomorphic to the groups $\underline{\hspace{3cm}}$ and $\underline{\hspace{3cm}}$.

The following is known as the *Jordan–Hölder* theorem:

THEOREM. Any two composition series of one and the same group are isomorphic.

DEFINITION. *The cyclic group of finite order n* (denoted by C_n) is the group consisting of elements $e, g, g^2, \ldots, g^{n-1}$ with multiplication subject to the condition $g^n = e$. (Here, of course, $g^2 = g \cdot g, g^3 = g \cdot g \cdot g$, etc.)

THEOREM. Every group of prime order p is cyclic.

DEFINITION. A group G is called *solvable* if it has a composition series in which each factor is a cyclic group whose order is a prime number p.

In many texts a group G is called solvable if there is a sequence

$$G \rhd G_1 \rhd \cdots \rhd G_r = \{e\}$$

such that G_i/G_{i+1} is Abelian. This definition can be shown to be equivalent to the one given here.

We have the following theorems [A, p. 70]:

THEOREM. Any subgroup of a solvable group is solvable.

THEOREM. The homomorphic image of a solvable group is solvable.

THEOREM. All Abelian groups are solvable.

1.2 PERMUTATIONS AND PERMUTATION GROUPS

DEFINITION. A *permutation* of a set S is a one-to-one mapping of S onto S.

If S contains n elements, we may think of the elements as numbered from 1 to n and of the permutation π as a mapping of the set $\{1, 2, \ldots, n\}$ onto $\{1, 2, \ldots, n\}$.

DEFINITION. A *cycle* $(ijk\cdots mn)$ is a permutation π such that $\pi(i) = j, \pi(j) = k, \ldots,$ $\pi(m) = n$ and $\pi(n) = i$. A cycle containing one element only, say (i), indicates that this element remains fixed. Two cycles are called *disjoint* if they act on disjoint sets; that is, if $\pi_1 = (i \cdots n)$ and $\pi_2 = (h \cdots l)$, then $\{i, \ldots, n\} \cap \{h, \ldots, l\} = \varnothing$.

Multiplication of two permutations means carrying out one after the other. As with functional notation in general, there is no universal agreement whether $\pi_1\pi_2$ means that π_1 or π_2 is to be carried out first. We shall generally assume that the transformation or permutation next to the argument is applied first, so that the associative law can be used. Thus, for instance, $(\pi_1\pi_2)(x) = \pi_1(\pi_2(x))$, and π_2 is

1.2 PERMUTATIONS AND PERMUTATION GROUPS

applied first in this case. In any case of possible ambiguity, the immediate context and/or use of parentheses should make it clear what is meant. In the multiplication tables for noncommutative groups and groups of permutations we shall also always indicate in the context which factor is to be applied first. This allows us to choose the notation best suited to the context.

It is easily seen (1) that every permutation is a product of disjoint cycles; (2) disjoint cycles commute with each other; (3) multiplication of permutations is associative, because it is really just functional composition which is surely associative; (4) there is an identity permutation, denoted (1), which leaves every element fixed; and (5) every permutation has an inverse. If a permutation is written as a product of disjoint cycles, it is understood that any element not mentioned explicitly remains fixed. Cycles of length 1 (containing one element only) are therefore generally omitted.

Every permutation of $\{1, 2, \ldots, n\}$ can be represented by a table of values. For instance, the permutations $\pi_1 = (143)(27)$ and $\pi_2 = (26)(345)$ on the set $S = \{1, \ldots, 7\}$ correspond to the tables

x	1	2	3	4	5	6	7
$\pi_1(x)$	4	7	1	3	5	6	2

and

x	1	2	3	4	5	6	7
$\pi_2(x)$							

From this we get the new tables

x	1	2	3	4	5	6	7
$(\pi_1\pi_2)(x)$	4	6	3	5	1	7	2

x	1	2	3	4	5	6	7
$(\pi_2\pi_1)(x)$							

where $(\pi_i\pi_j)(x) = \pi_i(\pi_j(x))$. This shows that in general multiplication of permutations is not commutative. With the understanding that the permutation on the right is applied first, the product permutations can be written using disjoint cycles in the form

$$\pi_1\pi_2 = (145)(267),$$

$$\pi_2\pi_1 = \underline{\qquad\qquad}.$$

DEFINITION.   The set $\mathfrak{S}_n$ consisting of the $n!$ possible permutations of $1, 2, \ldots, n$ forms a group called the *symmetric group* on n variables.

Any set of permutations that form a group is called a *permutation group*.

EXAMPLE.   Let us consider S_3, the symmetric group on three elements. Written in the cyclic notation for permutations, we can list its elements as follows: $S_3 = \{(1),$ $(12), \underline{\qquad}, \underline{\qquad}, \underline{\qquad}, \underline{\qquad}\}$, so that $|S_3| = \underline{\qquad}$,

$H_1 = \{(1),(12)\}, \qquad\qquad |H_1| = \underline{\qquad}, \ \mathrm{ind}_G H_1 = \underline{\qquad}$

$H_2 = \{\underline{\qquad\qquad}\}, |H_2| = \underline{\qquad}, \ \mathrm{ind}_G H_2 = \underline{\qquad}$

$H_3 = \{\underline{\qquad\qquad}\}, |H_3| = \underline{\qquad}, \ \mathrm{ind}_G H_3 = \underline{\qquad}$

$K = \{(1),(123),(132)\}, \qquad |K| = \underline{\qquad}, \ \mathrm{ind}_G K = \underline{\qquad}$

Partially for practice, and partially because this group will come up again later, we list in Table 1.2 the various right and left cosets of some of the subgroups. You will see that for some elements a, we have $aH \neq Ha$.

1.2 PERMUTATIONS AND PERMUTATION GROUPS

TABLE 1.2

Right cosets:	(1)	(12)	(13)	(23)	(123)	(132)	Left cosets:	(1)	(12)	(13)	(23)	(123)	(132)
$H_1(1)$	✓	✓					$(1)H_1$						
$H_1(12)$							$(12)H_1$						
$H_1(13)$													
$H_1(23)$													
$H_1(123)$													
$H_1(132)$													
$K(1)$	✓				✓	✓	$(1)K$						
$K(12)$							$(12)K$						
$K(13)$													
$K(23)$													
$K(123)$													
$K(132)$													

Returning to the concept of a solvable group, we shall see that $\mathfrak{S}_2$, $\mathfrak{S}_3$, and $\mathfrak{S}_4$ are solvable, but we have the

THEOREM. The symmetric group $\mathfrak{S}_n$ on n letters is not solvable for $n \geq 5$.

Proof. (1) If $\mathfrak{S}_n$ were solvable, then there would be a sequence

$$\mathfrak{S}_n = G_0 \rhd G_1 \rhd \cdots \rhd G_r = (1)$$

in which each G_{s+1} is a normal subgroup of G_s and G_s/G_{s+1} is Abelian (in fact, even cyclic).

(2) We shall show by induction on s that each G_s must contain every 3-cycle (ijk) for any i, j, k between 1 and n.

(3) Taking $s = 0$, we have $G_0 = \mathfrak{S}_n$, and $\mathfrak{S}_n$ surely contains all 3-cycles.

(4) Supposing that G_s contains every 3-cycle, we shall show that G_{s+1} also does.

(5) Let $\pi \in G_s$ and let φ be the natural homomorphism of G_s onto G_s/G_{s+1}, that is, $\varphi : \pi \rightarrow (\pi G_{s+1})$.

(6) Pick i, j, k, l, m to be any five distinct numbers $\leq n$. Since $n \geq 5$, this is possible.

(7) Then the 3-cycles $\pi_1 = (mji)$ and $\pi_2 = (ilk)$ are in G_s.

(8) By hypothesis, G_s/G_{s+1} is Abelian, so

$$\varphi(\pi_1^{-1}\pi_2^{-1}\pi_1\pi_2) = \varphi(\pi_1^{-1})\varphi(\pi_2^{-1})\varphi(\pi_1)\varphi(\pi_2)$$

$$= \varphi(\pi_1^{-1})\varphi(\pi_1)\varphi(\pi_2^{-1})\varphi(\pi_2)$$

$$= \varphi(\pi_1^{-1}\pi_1\pi_2^{-1}\pi_2)$$

$$= \varphi(1) = G_{s+1}, \quad \text{the identity element of } G_s/G_{s+1},$$

that is, $\pi_1^{-1}\pi_2^{-1}\pi_1\pi_2 \in G_{s+1}$.

(9) But $\pi_1^{-1}\pi_2^{-1}\pi_1\pi_2 = (ijm)(kli)(mji)(ilk) = (ijk)$, so we have $(ijk) \in G_{s+1}$.

(10) Since i, j, k were any integers $\leq n$, this means that all 3-cycles are in G_{s+1}.

(11) But then surely we can never have $G_r = (1)$, which proves the theorem. ‖

For $\mathfrak{S}_2$, $\mathfrak{S}_3$, and $\mathfrak{S}_4$ we have the following composition series:

$$\mathfrak{S}_2 \underset{\text{max}}{\triangleright} \{(1)\},$$

$$\mathfrak{S}_3 \underset{\text{max}}{\triangleright} \{(1), (123), (132)\} \underset{\text{max}}{\triangleright} \{(1)\},$$

$$\mathfrak{S}_4 \underset{\text{max}}{\triangleright} \underline{\hspace{2cm}} \underset{\text{max}}{\triangleright} \mathfrak{B}_4 \underset{\text{max}}{\triangleright} C_2 \underset{\text{max}}{\triangleright} \{(1)\},$$

where

$$\mathfrak{A}_4 = \{(1), (123), (132), (234), (243), (134), (143),$$
$$(124), (142), (12)(34), (13)(24), (14)(23)\},$$

$$\mathfrak{B}_4 = \{(1), (12)(34), (13)(24), (14)(23)\},$$

$$C_2 = \{(1), (12)(34)\}.$$

which shows that these groups are solvable.

We shall also come across the following concept:

DEFINITION. A permutation group G on $1, \ldots, n$ is called *transitive* if for any $1 \le k \le n$ it contains a permutation π which sends 1 into k.

DEFINITION. The *alternating group* $\mathfrak{A}_n$ is the group consisting of all those permutations on the subscripts of $x_1, \ldots, x_n$ which carry the function $\prod_{1 \le i < j \le n} (x_i - x_j)$ into itself.

1.3 FIELDS

DEFINITION. A *field* $\mathscr{F} = \langle F, +, \cdot, 0, 1 \rangle$ is a set of elements on which two binary operations $+$ and $\cdot$ are defined and containing two distinguished elements 0 and 1 with the properties that

(1) $\langle F, +, 0 \rangle = F^+$ and $\langle F - \{0\}, \cdot, 1 \rangle = F^\times$ are Abelian groups.

(2) The distributive law holds:

$$x(y + z) = xy + xz \qquad \text{for all } x, y, z \in F.$$

As with groups, it generally causes no confusion to denote a field simply by F, rather than $\langle F, +, \cdot, 0, 1 \rangle$.

DEFINITION. Two fields F_1 and F_2 are *isomorphic* if there is a mapping φ which maps F_1 one to one onto F_2 and if for every $a, b \in F_1$ we have $\varphi(a + b) = \varphi(a) + \varphi(b)$ and $\varphi(ab) = \varphi(a) \cdot \varphi(b)$. If $F_1 = F_2$, then φ is called an *automorphism*.

We shall use the following

NOTATION

Q for the field of rationals.

R for the field of reals.

$\mathscr{C}$ for the field of complex rationals: $\mathscr{C} = \{a + bi | a, b \in Q\}$.

C for the field of complex numbers: $C = \{a + bi | a, b \in R\}$.

A for the algebraic closure of the rationals.

$F(x)$ for the field of all formal quotients $p(x)/q(x)$, where $p(x), q(x)$ are formal polynomials in the indeterminate x, $q(x) \neq 0$, and we use the usual rules of addition, multiplication, and cancellation.

$F(\alpha)$ for the smallest subfield containing both F and α of the algebraic closure of F, where α is a root of some irreducible polynomial with coefficients in F.

$F(\alpha, \beta)$ for the smallest subfield containing F, α, and β of the algebraic closure of F, where α and β are roots of irreducible polynomials with coefficients in F.

ω for the primitive cube root of unity, $\omega = -\frac{1}{2} + \frac{1}{2}\sqrt{-3}$.

We see immediately that $F(\alpha, \beta) = F(\beta, \alpha)$, $\mathscr{C} = Q(i)$ and $Q(\omega) = Q(\sqrt{-3})$.

EXERCISES. (1) The fields $Q(\sqrt{2})$ and $Q(-\sqrt{2})$ are the same. Show that the correspondence

$$a + b\sqrt{2} \leftrightarrow a - b\sqrt{2}$$

is an automorphism. Can all elements of $Q(\sqrt{2})$ be written in the form $a + b\sqrt{2}$?

_____

(2) Is $Q(\sqrt{2}) \cong Q(\sqrt{-2})$? _____ Is the correspondence $a + b\sqrt{2} \leftrightarrow a + b\sqrt{-2}$ an isomorphism? _____ Why? _____

_____

1.4   RINGS AND POLYNOMIALS

(3) $Q(\sqrt{2}) \not\cong Q(\sqrt{3})$ because _____

_____

We shall assume familiarity with the usual rules of arithmetic in fields that follow from the axioms, as, for example: For every $a, b \in F$ we have $a \cdot 0 = 0$, and $(-a)(-b) = ab$.

DEFINITION.   (1) A field F is of finite *characteristic p* [char$(F) = p$] if there is a least positive integer p such that $\underbrace{(1 + 1 + \cdots + 1)}_{p \text{ times}} = 0$ in F.

(2) If there is no such integer, then F is said to be of *characteristic* 0.

Since $\langle F - \{0\}, \cdot, 1 \rangle = F^{\times}$ is always a group, there can be no divisors of zero in a field. From this we see immediately that the characteristic m must be a prime number (for if it were not, say $m = k \cdot l$, then_____

_____).

The simplest examples of fields of characteristic p (p a prime) are the fields Z_p with addition and multiplication carried out mod p. These are finite fields, but it is easy to form infinite fields of finite characteristic. For example, we can form $Z_p(x)$, the field of all rational functions in x with coefficients in Z_p. A few important theorems about finite fields are stated at the end of Section 1.4.

1.4   RINGS AND POLYNOMIALS

We include several more familiar definitions and basic theorems because we shall need them later. See also Table 1.3.

DEFINITION. $\langle R, +, \cdot, 0, 1 \rangle$ is a *commutative ring* if

(1) $\langle R, +, 0 \rangle = R^{+}$ is an Abelian group.

(2) $\langle R, \cdot \rangle$ satisfies the closure, associative, and commutative axioms.

(3) The distributive law holds: For every $a, b, c \in R : a \cdot (b + c) = a \cdot b + a \cdot c$.

TABLE 1.3 Examples of Rings

	R	$+$	$\cdot$	Is it a ring?	Is there a unit?	Is it commutative?	Is it an integral domain?	Is it a field?	Comments (list laws that fail)
1	Z	$+$	$\cdot$						
2	Z	$+$	$+$						
3	Z_3	$+_{(3)}$	$\cdot_{(3)}$						
4	Z_3	$\cdot_{(3)}$	$+_{(3)}$						
5	Z_4	$+_{(4)}$	$\cdot_{(4)}$						
6	Z_p, p prime	$+_{(p)}$	$\cdot_{(p)}$						
7	$2Z$	$+$	$\cdot$						
8	nZ	$+$	$\cdot$						
9	C	$+$	$\cdot$						
10	$\{a + b\sqrt{2}\}$	$+$	$\cdot$						
11	$\{a + b\sqrt{-2}\}$	$+$	$\cdot$						
12	$\mathfrak{S}(S)$	$\cap$	$\cup$						
13	$\mathfrak{S}(S)$	$\oplus$	$\cap$						
14	$\mathfrak{S}(S)$	$\oplus$	$\cup$						
15	V_3	$+$	$\times$ (cross product)						
16	V_3	$+$	$\cdot$ (dot product)						
17	Q	$+$	$\cdot$						
18	Functions continuous on $-1 \leq x \leq 1$	$+$	$\cdot$						
19	Functions $f(x)$ for which $f(0) = 0$	$+$	$\cdot$						

Note: In (12), (13), and (14), S is a set and $\mathfrak{S}(S)$ is the power set of S (the set of all subsets of S), and for $U, V \subseteq S$, we have $U \oplus V = (U \cup V) - (U \cap V)$. In (18) and (19), the operations $+$ and $\cdot$ are pointwise addition and multiplication. In (15) and (16), V_3 is Euclidean three-dimensional vector space. Also $a, b \in Q$.

1.4 RINGS AND POLYNOMIALS

DEFINITION. R is a *commutative ring with an identity* if in addition it contains a distinguished element 1 which acts as a multiplicative identity: $a \cdot 1 = a$, for every $a \in R$.

DEFINITION. If R is a ring, $a, b \in R$, and $a \cdot b = 0$, but $a \neq 0$, $b \neq 0$, then a and b are called *divisors of zero* or *zero divisors*.

DEFINITION. A commutative ring which has no zero divisors is called an *integral domain*.

DEFINITION. If R is a commutative ring, then $R[x] = \{p(x)|$, where $p(x)$ is an expression of the form $a_0 + \cdots + a_n x^n$, with $a_i \in R$, n a positive integer$\}$, together with the usual addition and multiplication of polynomials is also a ring and is called the *ring of polynomials* in x with coefficients in R, and $p(x)$ is called a *polynomial* over the ring R.

Note. It is important to notice that here $p(x)$ is a formal sum, defined simply to be an expression of a certain form and must not be confused with the polynomial function whose domain is the commutative ring R. The $+$ sign and the classical kth power x^k of the "indeterminate" x do not denote addition or multiplication by x^k but only serve the same purpose as the comma and position in vector notation. More specifically, one may think of a polynomial as a function on the set $\{0, 1, 2 \ldots\}$ with values in R subject to the condition that all but a finite number of these values vanish. In fact, the ring of polynomials over the commutative ring R is often defined as the subspace of elements of R^ω in which all but a finite number of components vanish and in which there is an appropriate addition and multiplication. The vector concept is, however, quite awkward for multiplication or when considering polynomials in several variables. It is assumed that all readers have some familiarity with polynomials and all should be aware that $R[x]$ is a vector space and an algebra over R.

DEFINITION. If $p(x) = a_0 + \cdots + a_n x^n \in F[x]$ and $a_n \neq 0$, then the *degree* of $p(x)$ is n [written $\deg p(x) = n$].

DEFINITION. The polynomial $p(x)$ is *irreducible* over F if there are no polynomials $q(x)$ and $r(x) \in F[x]$ of degree ≥ 1 such that $p(x) = q(x) \cdot r(x)$.

EXAMPLE. Let $p(x) = x^2 + 1$. Then $p(x)$ is irreducible over R, but it factors into

$$p(x) = (x + i)(\underline{\hspace{2cm}}) \text{ over } \mathscr{C}$$

$$= (x + 1)(\underline{\hspace{2cm}}) \text{ over } Z_2$$

$$= (x + 2)(\underline{\hspace{2cm}}) \text{ over } Z_5.$$

DEFINITION. If $R_1 \subseteq R_2$ and R_1 and R_2 are both rings with the same operations, then R_1 is called a *subring* of R_2. We shall write $R_1 < R_2$.

DEFINITION. If $I < R$ and for every $a \in R$ we have $aI \subseteq I$ and $Ia \subseteq I$, then I is called a two-sided *ideal in R*.

DEFINITION. If I is an ideal in R, then we say $a \equiv b \bmod I$ whenever $(a - b) \in I$.

It is an easy exercise to prove that this is an equivalence relation.

EXAMPLES. (1) Let $R_2 = Z, R_1 = 3Z = \{3a | a \in Z\} = \{$all multiples of 3$\}$. Then $a \equiv b \bmod R_1 \Leftrightarrow a \equiv b \bmod 3$ (in the usual sense of congruence modulo an integer), and $\forall n \in Z : n \equiv 0$ or 1 or 2 mod 3.

(2) Let

$$R_2 = Q[x], R_1 = (x^2 + 2)R_2 = \{(x^2 + 2) \cdot q(x) \mid q(x) \in Q[x]\}$$

$$= \{\text{all multiples of } (x^2 + 2)\}.$$

We then have $p(x) \in Q[x]; p(x) \equiv a + bx \bmod R_1$, where a and b are rationals. In addition, it is easily seen that $a_1 + b_1 x \equiv a_2 + b_2 x \bmod R_1$ iff $a_1 = a_2$ and $b_1 = b_2$.

1.4 RINGS AND POLYNOMIALS

DEFINITION. If the ideal I consists of all multiples of some element m of R, that is, if $I = \{ma|a \in R\}$, then I is called a *principal ideal*.

NOTATION. The principal ideal determined by m is denoted by (m). For example, if $p(x)$ is a polynomial with coefficients in the field F, then

$$(p(x)) = \{p(x) \cdot q(x) \mid q(x) \in F[x]\}.$$

Two-sided ideals play somewhat the same role in ring theory as do normal subgroups in group theory. They are used to form quotient rings just as normal subgroups are used to form quotient groups. Polynomial rings $F[x]$ will be used throughout the book, so we first give an example involving polynomial rings before stating the general definition of quotient rings.

The ring $F[x]$ is an Abelian group under $+$ and the ideal $(p(x))$ is a normal subgroup so we can form the quotient group $F[x]/(p(x))$, using the definition in Section 1.1 with $H = (p(x))$ and $G = F[x]$, and $*$ as $\cdot$. The elements of this quotient group are the congruence classes of $F[x]$ taken modulo $p(x)$. Let $[r(x)]$ denote the congruence class of $r(x) \in F[x]$; that is, $[r(x)] = \{q(x)|r(x) \equiv q(x) \bmod(p(x))\}$. Addition of congruence classes is automatically defined by the definition of a quotient group, and we can define multiplication analogously:

$$[r_1(x)] + [r_2(x)] = [r_1(x) + r_2(x)],$$

$$[r_1(x)] \cdot [r_2(x)] = [r_1(x) \cdot r_2(x)].$$

EXAMPLE. If $p(x) = x^2 + 2$, we have

$$[2x + 3] + [x - 1] = [3x + 2],$$

$$[2x + 3] \cdot [x - 1] = [2x^2 + x - 3] = [x - 7],$$

because

$$2x^2 + x - 3 = (\underline{\quad\quad})(x^2 + 2) + (\underline{\quad\quad\quad})$$

$$\equiv (\underline{\quad\quad\quad}) \bmod(x^2 + 2).$$

Is this multiplication of congruence classes associative? _____ Commutative? _____ Is there a multiplicative identity element? _____ Is it distributive over the addition of congruence classes? _____

From the answers to these questions, it is easily seen that the congruence classes themselves form a ring, which is called the quotient ring of $F[x]$ by $(p(x))$ and is denoted by $F[x]/(p(x))$. In practice, the brackets are often omitted around the elements representing the classes, but computations must be carried out mod $p(x)$. More generally, we make the following definition:

DEFINITION. If R is a ring and I is an ideal in R, then the *quotient ring* R/I is the algebraic system $\langle R/I, \oplus, \odot \rangle$, where R/I is the set of distinct cosets $a + I$ and the addition $\oplus$ and multiplication of these cosets is defined by

$$(a + I) \oplus (b + I) = (a + b) + I,$$

$$(a + I) \odot (b + I) = ab + I.$$

(Using the notation $[a]$ for the coset containing a, we write the last equations in the form $[a] \oplus [b] = [a + b], [a] \odot [b] = [ab]$. In general we shall write $+, \cdot$ for $\oplus, \odot$ in the quotient ring, and assume that there is no confusion.)

It is of course necessary to check that these operations are well defined and that R/I is really a ring.

EXAMPLE.   Let

$R = \{f | f$ is a continuous real function of one variable x for
$\quad -1 \le x \le +1\} = C([-1, +1]),$

$I_c = \{g | g \in R$ and $g(c) = 0$, where c is some fixed real
$\quad$ number between -1 and $+1\}.$

(I_c is called the maximal ideal fixed at the point c.) Then for any $f \in R$, we have

$$[f] = f + I_c = \{f + g \mid g(c) = 0\} = \{h | h(c) = f(c)\}.$$

1.4 RINGS AND POLYNOMIALS

If $f(c) = r\,(r \in R)$, then the correspondence $[f] \leftrightarrow r$ is one to one between the elements of R/I_c and R. In fact, it is an isomorphism, because _____

_____. In conclusion, therefore, $R/I_c \cong$ R.

In this example, is $I_{c_1} \cup I_{c_2}$ an ideal _____? Is $I_{c_1} \cap I_{c_2}$? _____ Can you list all the ideals of R?

DEFINITION.   If I is an ideal in a ring R, then I is called *maximal* if for every ideal $J : I \subseteq J \subseteq R$ implies that $J = I$ or $J = R$.

What are the maximal ideals of the ring in the above example? _____

_____

Let R^2 be the Cartesian product of R with itself with $+$ and $\cdot$ defined component-wise. Then we can show that $R/(I_{c1} \cap I_{c2}) \cong R^2$ by checking that the correspondence

$$[f] \leftrightarrow \langle \underline{\quad\quad}, \underline{\quad\quad} \rangle$$

is an isomorphism.

A very useful theorem on rings states:

THEOREM.   If R is a commutative ring with an identity element 1 and I is a maximal ideal of R, then R/I is a field. Conversely also if R/I is a field, then I is maximal in R.

We may apply this theorem to polynomial rings $F[x]$, where F is a field. We have

$$(p_1(x) \cdot p_2(x)) = (p_1(x)) \cap (p_2(x)),$$

so if the ideal $(p(x))$ is maximal, then $p(x)$ must be irreducible, that is, have no factors in $F[x]$. Conversely, if $p(x)$ is irreducible, then the ideal $(p(x))$ is maximal: For if $q(x)$ is any polynomial not in $(p(x))$, then $\gcd(q(x), p(x)) = 1$, so we can find polynomials $a(x)$ and $b(x)$ such that $a(x)p(x) + b(x)q(x) = 1$; hence adjoining $q(x)$ to $(p(x))$ and forming a new ideal I implies that $1 \in I$, hence I is the whole ring. So $(p(x))$ is maximal. We can, therefore, apply the theorem above to conclude that $F[x]/(p(x))$ is a field whenever $p(x)$ is irreducible. The zero element of this field is $[p(x)]$.

As another application, suppose that

$$R = Z[i] \text{ and } I = (1 + i) = \{Z(1 + i) \mid Z \in R\}.$$

Then $R/I \cong Z_2$, because _____

_____. Suppose now that $R = Z[i]$ and $I = (2 + i)$. Will R/I

again be a field? _____ Is $Z[i]/(2)$ a field? _____

Suppose now that $F = Z_p$, where p is a prime. One can show that for every

prime p and for every integer $n \geq 1$ there is a polynomial $f(x)$ of degree n that is

irreducible over F. The ideal $(f(x))$ is then maximal in $F[x]$ and $F[x]/(f(x))$ is a field.

It will be of characteristic p and contain p^n elements, because there are exactly p^n pos-

sible polynomials $a_0 + \cdots + a_{n-1}x^{n-1}$ of degree $\leq n - 1$ with coefficients in F and

each of these determines precisely one congruence class. In contrast to the situation

when $F = Q$ we have the following.

THEOREM. If $F = Z_p$ and $f(x), g(x)$ are any two polynomials of the same degree

irreducible over F, then $F[x]/(f(x)) \cong F[x]/(g(x))$.

All fields containing p^n elements are therefore isomorphic, which justifies the

DEFINITION. The *Galois field with p^n elements* $(GF(p^n))$ is the field $Z_p[x]/(f(x))$,

where $f(x)$ is any polynomial of degree n irreducible over p.

We end this section with the following somewhat surprising

THEOREM. The nonzero elements of $GF(p^n)$ form a cyclic group under multipli-

cation [vdW, p. 117].

These and other theorems about finite fields are proved in Section 4.11. As they

do, however, appear in some of the earlier examples, they are stated here for reference.

1.5 SOME ELEMENTARY THEORY OF EQUATIONS

Many of the examples we work with require a little facility in arithmetic with complex numbers, and also make use of some of the elementary theorems on the theory of equations. As customary, we let $i = \sqrt{-1}$ and write the complex number α as $\alpha = a + bi$, with a, b real.

DEFINITION. A complex number α is called *algebraic* if there is a polynomial $p(x)$ with rational coefficients which has α as a root.

DEFINITION. A complex number α is *expressible in terms of radicals* if there is a sequence of expressions $\beta_1, \beta_2, \ldots, \beta_n$, where β_1 is rational and each succeeding β_i is either also rational or else the sum, difference, product, quotient, or kth root of preceding β's, and the last β_n is α.

We have the following results:

(1) α is expressible in terms of radicals $\Rightarrow \alpha$ is algebraic,

(2) a, b algebraic $\Leftrightarrow \alpha = a + bi$ is algebraic,

(3) a, b are expressible in terms of radicals $\Leftrightarrow \alpha = a + bi$ so expressible.

(4) $\exists \alpha : \alpha$ is algebraic, but not expressible in terms of radicals.

(5) $\exists \alpha : \alpha$ is real and expressible in terms of radicals, but a, b are not expressible in terms of real radicals.

Remarks on Proofs of These. (1) will incidentally be proved later in these notes. It is not at all difficult.

(2) follows from a well-known theorem to the effect that if α, β are algebraic, then so are $\alpha \pm \beta$, $\alpha \cdot \beta$, and α/β (if $\beta \neq 0$). Since α and its conjugate $\bar{\alpha}$ are both roots of the same $p(x)$, so $\frac{1}{2}(\alpha + \bar{\alpha}) = a$ is algebraic.

(3) is trivial.

(4) will follow once we show that there are equations which are not solvable by radicals.

(5) is a result that arises out of the theory of cubic equations [vdW, p. 179].

EXAMPLES.   Insert a and b for the given α:

α	a	b
$\sqrt{i}$		
ω^2		
$\sqrt{\omega}$		
$\sqrt[3]{i}$		

(Here $\omega = -\frac{1}{2} + \frac{1}{2}\sqrt{-3}$.) Now for some more results from the theory of equations.

(6) If $a(x)$, $b(x)$ are polynomials with coefficients in a field F, and $b(x) \neq 0$, then we can find uniquely determined polynomials $q(x)$ and $r(x)$ such that

$$a(x) = b(x)q(x) + r(x),$$

where $r(x) = 0$ or else the degree of $r(x)$ is strictly less than the degree of $b(x)$.

(7) Any two polynomials $a(x)$ and $b(x)$ ($\neq 0$) over F have a gcd $c(x)$ and one can exhibit $p(x)$ and $q(x)$ such that

$$a(x)p(x) + b(x)q(x) = c(x).$$

(8) REMAINDER THEOREM. The remainder obtained on dividing $p(x)$ by $(x - \alpha)$ is $p(\alpha)$.

(9) FACTOR THEOREM. $(x - \alpha)$ is a divisor of $p(x) \Leftrightarrow p(\alpha) = 0$.

(10) FUNDAMENTAL THEOREM OF ALGEBRA. If $p(x)$ is a polynomial with rational (or even real or complex) coefficients, then there is a complex number α such that $p(\alpha) = 0$.

1.5 SOME ELEMENTARY THEORY OF EQUATIONS

(11) A polynomial $p(x)$ of degree n over the rationals factors uniquely into linear factors over the field of complex numbers, that is, $p(x) = a(x - \alpha_1) \cdots (x - \alpha_n)$, where $\alpha_1, \ldots, \alpha_n$ are the roots of $p(x)$; moreover, if

$$p(x) = a(x - \alpha_1)(x - \alpha_2) \cdots (x - \alpha_n)$$

$$= a(x^n - \sigma_1 x^{n-1} + \sigma_2 x^{n-2} \cdots + (-1)^{n-1} \sigma_n),$$

then

$$\sigma_1 = \sum_i \alpha_i, \sigma_2 = \sum_{i<j} \alpha_i, \alpha_j, \ldots, \sigma_n = \alpha_1 \alpha_2 \cdots \alpha_n.$$

DEFINITION. The *elementary symmetric functions* of $\alpha_1, \ldots, \alpha_n$ are defined to be

$$\sigma_1 = \sum_{1 \le i \le n} \alpha_i = \alpha_1 + \cdots + \alpha_n,$$

$$\sigma_2 = \sum_{1 \le i_1 < i_2 \le n} \alpha_{i_1} \alpha_{i_2} = \alpha_1 \alpha_2 + \alpha_1 \alpha_3 + \cdots + \alpha_{n-1} \alpha_n,$$

$$\sigma_r = \sum_{1 \le i_1 < \cdots < i_r \le n} \alpha_{i_1} \cdots \alpha_{i_r}, \qquad \text{for } 1 \le r \le n,$$

$$\sigma_n = \alpha_1 \cdots \alpha_n,$$

$$\sigma_r = 0, \text{ for } r > n.$$

DEFINITION. A polynomial $p(\alpha_1, \ldots, \alpha_n)$ is called *symmetric* in $\alpha_1, \ldots, \alpha_n$ if it remains unchanged by any permutation of $\alpha_1, \ldots, \alpha_n$.

(12) FUNDAMENTAL THEOREM ON SYMMETRIC FUNCTIONS. Every polynomial $p(\alpha_1, \ldots, \alpha_n)$ symmetric in $\alpha_1, \ldots, \alpha_n$ can be expressed in a unique manner as a polynomial in the elementary symmetric functions of $\alpha_1, \ldots, \alpha_n$.

EXAMPLES

(1) $p(\alpha_1, \ldots, \alpha_n) = \sum_i \alpha_i^2 = \sigma_1^2 - 2\sigma_1$.

(2) $p(\alpha_1, \ldots, \alpha_n) = \sum_i \alpha_i^3 = \sigma_1^3 - 3\sigma_1\sigma_2 + 3\sigma_3$.

(3) $p(a_1, \ldots, \alpha_n) = \sum_{i \ne j} \alpha_i \alpha_j^2 = \sigma_1 \sigma_2 - 3\sigma_3$.

These can be checked by expanding the right-hand side. There is a systematic method for calculating the necessary polynomial of the elementary symmetric functions for any given symmetric polynomial $p(\alpha_1, \ldots, \alpha_n)$. An excellent exposition of this will be found in Weisner's theory of equations [W].

DEFINITION.   Let $s_k(\alpha_1, \ldots, \alpha_n) = \alpha_1^k + \cdots + \alpha_n^k$, where k is any integer (s_k is therefore symmetric in $\alpha_1, \ldots, \alpha_n$).

(13) NEWTON'S IDENTITIES. These can be used to evaluate s_k from the corresponding σ's.

$$s_1 - \sigma_1 = 0,$$

$$s_2 - \sigma_1 s_1 + 2\sigma_2 = 0,$$

$$s_3 - \sigma_1 s_2 + \sigma_2 s_1 - 3\sigma_3 = 0,$$

$$\vdots$$

$$s_k - \sigma_1 s_{k-1} + \sigma_2 s_{k-2} - \cdots + (-1)^k k\sigma_k = 0.$$

EXAMPLE.   Let $\alpha_1, \alpha_2, \alpha_3$ be the roots of $(x^3 + 2x^2 + 4x + 8)$. Then

$$\sigma_1 = -2,$$

$$\sigma_2 = \underline{},$$

$$\sigma_3 = \underline{},$$

and

$$s_4 - \sigma_1 s_3 + \sigma_2 s_2 - \sigma_3 s_1 = 0,$$

$$s_3 - \sigma_1 s_2 + \sigma_2 s_1 - 3\sigma_3 = 0,$$

$$s_2 - \sigma_1 s_1 + 2\sigma_2 = 0,$$

$$s_1 - 1 \cdot \sigma_1 = 0.$$

1.5 SOME ELEMENTARY THEORY OF EQUATIONS

Substituting the known values of $\sigma_1, \sigma_2, \sigma_3$ we get

$$s_4 + 2s_3 + 4s_2 + 8s_1 \quad = 0,$$

$$s_3 + 2s_2 + 4s_1 - \underline{\hphantom{xxxx}} = 0,$$

$$\underline{\hphantom{xxxxxxxxxxxx}} = 0,$$

$$s_1 - \underline{\hphantom{xxxx}} = 0.$$

The solution of this system of equations is $s_1 = \underline{\hphantom{xxx}}$, $s_2 = \underline{\hphantom{xxx}}$, $s_3 = \underline{\hphantom{xxx}}$, $s_4 = \underline{\hphantom{xxx}}$.

(The result should not be too surprising, because $\alpha_1, \alpha_2, \alpha_3$, are different values of $\sqrt[4]{16}$.)

(14) If $\bar{\alpha}$ is the complex conjugate of α and $p(x)$ any polynomial over Q (or R), then $p(\bar{\alpha}) = \overline{p(\alpha)}$. Hence if $p(\alpha) = 0$, then $p(\bar{\alpha}) = \bar{0} = 0$, so that the complex roots of polynomials with real coefficients can always be paired off into conjugates.

(15) If $p(x) = c_n x^n + \cdots + c_0$ is a polynomial with rational coefficients and $\alpha = a/b$ is a root, where a, b are relatively prime integers, then b is a divisor of c_n and a is a divisor of c_0. (We write $b|c_n$ and $a|c_0$.)

This easy result is very useful, because it reduces the problem of finding all rational roots (if any) of a given polynomial to a finite number of trials.

EXAMPLE. Given $p(x) = 12x^3 + 16x^2 - 7x - 6$. Here $c_n = 12$, and $b|c_n$, so the possibilities are $b = \pm1, \pm2, \pm3, \pm4, \pm6, \pm12$, while $c_0 = -6$, and $a|c_0$, so the possibilities are $a = \pm1, \pm2, \pm3, \pm6$. The possible roots a/b are therefore $a/b = \pm1, \pm\frac{1}{2}, \pm\frac{1}{3}, \pm\frac{1}{6}, \pm2, \pm\frac{2}{3},$ _____

_____

Substituting these into the original equation we find that _____, _____, _____, are, in fact, roots.

(16) THEOREM (Gauss). If the polynomial $f(x)$ has integer coefficients and can

be factored into factors with rational coefficients, then $f(x)$ actually has factors all having integer coefficients only.

EXAMPLE

$$f(x) = 12x^3 + 16x^2 - 7x - 6 = (3x + \tfrac{9}{2})(4x^2 - \tfrac{2}{3}x - \tfrac{4}{3})$$

$$= \underline{\hspace{8cm}}.$$

(17) EISENSTEIN IRREDUCIBILITY CRITERION. If $f(x) = c_0 + c_1 x + \cdots + c_n x^n$ is a polynomial with integer coefficients and there is a prime p such that

(1) $p|c_0, p|c_1, \ldots, p|c_{n-1}$,

(2) $p \nmid c_n$ (p does not divide c_n),

(3) $p^2 \nmid c_0$,

then $f(x)$ is irreducible over the rationals.

(The converse of this theorem is not true.)

EXAMPLE. (1) Let $f(x) = x^4 + 4x^2 + 4x + 2$ and let $p = \underline{\hspace{1.5cm}}$. This polynomial is therefore irreducible over Q.

(2) Let $f(x) = 2x^4 + 4x^2 + 4x + 1$; substitute $x = 1/y$ and then let $p = \underline{\hspace{1.5cm}}$. This polynomial is $\underline{\hspace{3cm}}$.

(3) Let $f(x) = x^4 + x^3 + x^2 + 1$; substitute $x = 1 + y$, to get $f(1 + y) = \underline{\hspace{6cm}}$ and now take $p = \underline{\hspace{1.5cm}}$. We see that $f(1 + y)$ is reducible/irreducible (cross out one) and therefore so is $f(x)$.

(18) There is actually an algorithm for deciding whether every given polynomial with rational coefficients factors over the rationals. The procedure is due to Kronecker [vdW, p. 77].

(19) Incidentally, every polynomial with *real* coefficients factors into linear and irreducible quadratic factors over the reals. The cofficients of these factors are not always expressible in terms of radicals, as we shall see. Sometimes, however, they are and are not difficult to find.

1.5 SOME ELEMENTARY THEORY OF EQUATIONS

EXAMPLES.   (1) $x^2 + 1$ is irreducible over the reals R.

But:

(2) $x^4 + 1 = (x^2 + \sqrt{2}x + 1)(x^2 + $ _____) over R.

(3) $x^8 + 1 = (x^2 + \sqrt{2 + \sqrt{2}}x + 1)($_____)

_____) (_____) over R.

(20) THEOREM (Kronecker). If F is any field and $f(x) \in F[x]$, then there exists an extension K of F in which $f(x)$ has a root α.

Proof.   If $p(x)$ is an irreducible factor of $f(x)$, then $F[x]/(p(x))$ is a field that contains a root of $p(x)$—the coset $x + (p(x))$—and a subfield F' isomorphic to F—the field F' consisting of all the cosets $a + (p(x))$, with $a \in F$. If we let α be any symbol and carry out addition and multiplication modulo $p(\alpha)$, then the elements of $F[\alpha]$ will form a field K and $K \cong F[x]/(p(x))$ by the isomorphism $\alpha \leftrightarrow x + (p(x))$. Clearly $F < K$ and since $p(x)$ is a factor of $f(x)$ we have $f(\alpha) \equiv 0 \bmod p(\alpha)$, so the theorem is proved. [The field K so formed is generally denoted by $F(\alpha)$.]   ‖

(21) THEOREM. For any field F we have $(F(\alpha))(\beta) \cong (F(\beta))(\alpha)$.

Proof.   Exercise.

NOTATION.   We usually write $F(\alpha, \beta)$ instead of $(F(\alpha))(\beta)$. The theorem can then be stated as

$$F(\alpha, \beta) = F(\beta, \alpha).$$

(22) DEFINITION. The field E is called a *simple* extension of F if there is an element $\theta \in E$ such that $E = F(\theta)$. The following beautiful theorem is due to Steinitz:

THEOREM.   If $[E:F]$ is finite, then E is a simple extension of F if and only if there are only a finite number of intermediate fields.

Proof.   See [A, p. 64].

We have the following corollary:

COROLLARY. If F is a finite field or is of characteristic 0 and $E = F(\alpha_1, \ldots, \alpha_n)$, then E contains an element θ such that $E = F(\theta)$, provided $[E:F]$ is finite.

EXAMPLE. Let $F = Q$, $\alpha_1 = i$, $\alpha_2 = \sqrt{2}$, $\theta = i + \sqrt{2}$, $E = F(\alpha_1, \alpha_2)$. We wish to show that $E = F(\theta) = Q(i + \sqrt{2})$. We have

$$\frac{\theta^2 - 3}{2\theta} = \underline{\hspace{3cm}} = i,$$

$$\frac{\theta^2 + 3}{2\theta} = \underline{\hspace{3cm}} = \sqrt{2},$$

which shows that i and $\sqrt{2}$ are in $F(\theta)$. Therefore, $E < F(\theta)$. But clearly $\theta \in E$, so $F(\theta) < E$, too. Therefore, $E = F(\theta)$.

Before we go any further in the study of fields there is a warning: Don't underestimate the difficulties involved in mere "straightforward" arithmetic. As examples, take the following:

EXERCISES. (1) Show that $\sqrt[3]{\frac{7}{2} + \frac{7}{18}\sqrt{-3}} + \sqrt[3]{\frac{7}{2} - \frac{7}{18}\sqrt{-3}}$ is a real number. (You can do this by showing that it is a root of the equation $\underline{\hspace{3cm}}$, which has only real roots, because $\underline{\hspace{6cm}}$

$\underline{\hspace{7cm}}$.)

(2) Simplify $\sqrt{5 + \sqrt{24}}$.

(3) Simplify $\sqrt[3]{1 + \frac{2}{3}\sqrt{\frac{7}{3}}} + \sqrt[3]{1 - \frac{2}{3}\sqrt{\frac{7}{3}}}$ [W, p. 179] by showing that it is the only real root of the equation $\underline{\hspace{3cm}}$ which factors

$\underline{\hspace{4cm}}$.

(4) Construct a more complicated example.

1.6 VECTOR SPACES

Much of the theory of algebraic extensions can be very elegantly treated by thinking of fields as vector spaces over their various subfields, so we list some of the

1.6 VECTOR SPACES

definitions and theorems that will be needed and include examples. See also Table 1.4.

DEFINITION. An algebraic system $\langle V, \oplus, \cdot, F \rangle$ is a *vector space over the field F* if

(1) $\langle V, \oplus \rangle$ is an Abelian group, and

(2) $\forall a, b \in F$ and $x, y \in V$ we have

$$(a + b)x = a \cdot x \oplus b \cdot x,$$

$$a \cdot (x \oplus y) = a \cdot x \oplus a \cdot y,$$

$$a \cdot (b \cdot x) = (ab) \cdot x,$$

(3) $1 \cdot x = x$.

The elements of V are called *vectors*.

DEFINITION. W is a *subspace of V* (written $W < V$) if

(1) $W \subseteq V$, and

(2) $\langle W, \oplus, \cdot, F \rangle$ is also a vector space.

DEFINITION. The elements $v_1, \ldots, v_k$ of V are called *linearly independent* if for every $a_1, \ldots, a_k \in F$:

$$a_1 v_1 \oplus \cdots \oplus a_k v_k = \mathbf{0} \Rightarrow a_1 = \cdots = a_k = 0.$$

(Here $\mathbf{0}$ denotes the zero vector, whereas 0 is the zero element of F.)

The elements $v_1, \ldots, v_k \in V$ are called *linearly dependent* if they are not independent, that is, if there are $a_1, \ldots, a_k \in F$ such that

$$a_1 v_1 \oplus \cdots \oplus a_k v_k = \mathbf{0} \text{ and some } a_i \neq 0.$$

DEFINITION. The *dimension of V over F* ($\dim_F V$) is the maximum number of linearly independent elements in V.

DEFINITION. If $S \subseteq V$, then the *linear span of S, L(S)* is the vector space formed by the set of all finite linear combinations of elements of S.

TABLE 1.4 Examples of Vector Spaces

	V	F	Basis	$\dim_F V$
1	{polynomials in x of degree $\leq n$, with coefficients in Q}	Q	$\{1, x, \ldots, x^n\}$	$n + 1$
2	{formal infinite series in nonnegative powers of x with coefficients in Q}	Q		
3	{$(n \times n)$ matrices with elements in Q}	Q		
4	$\{a + b\sqrt{3}\,|\,a, b \in Q\}$	Q		
5	Same	$Q(\sqrt{3})$		
6	$\{a + b\sqrt{3}\,|\,a, b \in \mathsf{R}\}$	R		
7	$\{a + b\sqrt{-3}\,|\,a, b \in Q\}$	Q		
8	Same	$Q(\sqrt{-3})$		
9	$\{a + b\sqrt{-3}\,|\,a, b \in \mathsf{R}\}$	R		
10	$Q[x]/(x^3 - 2)$	Q		
11	$\{a + b\sqrt[3]{2}\,|\,a, b \in Q\}$	Q		
12	$\{a + b \cdot 2^{1/3} + c \cdot 2^{2/3}\,|\,a, b, c \in Q\}$	Q		
13	$\{a + b \cdot 2^{1/4}\,|\,a, b \in Q\}$	Q		
14	$Q(2^{1/4})$	Q		
15	$Q[x]/(x^4 - 2)$	Q		
16	$\{a + b\sqrt{2} + c\sqrt{3} + d\sqrt{6}\,|\,a, b, c, d \in Q\}$	Q		
17	Same	$Q(\sqrt{2})$		

1.6 VECTOR SPACES

DEFINITION. The set of elements $B = \{v_1, \ldots, v_n\}$ is a *basis* for V if

(1) $v_1, \ldots, v_n$ are linearly independent, and

(2) $L(B) = L(\{v_1, \ldots, v_n\}) = V$.

THEOREM. If $B_1 = \{v_1, \ldots, v_n\}$ and $B_2 = \{w_1, \ldots, w_m\}$ are both bases for V, then $n = m = \dim_F V[A]$.

THEOREM. If $\dim_F V = n$ and $W < V$, then $\dim_F W \leq n$.

CHAPTER II

FIELDS

2.1 DEGREE OF AN ALGEBRAIC EXTENSION

Let E, F be fields. If $F \subseteq E$, then E is called an *extension* of F, and F is called a *subfield* of E. We think of E as a vector space over F and construct a basis $B = \{\alpha_1, \alpha_2, \ldots\}$ in the usual inductive manner. Let $\alpha_1 = 1$, so that $\alpha_1 \in F$, and let $V_1 = V_F(\alpha_1)$, so in fact $V_1 = F$. Clearly $V_1 \subseteq E$. Next, if $V_1 \neq E$, let α_2 be any element of $(E - V_1)$ and let $V_2 = V_F(\alpha_1, \alpha_2)$. We may now have $V_2 = E$, but, if not, let α_3 be any element of $(E - V_2)$ and let $V_3 = V_F(\alpha_1, \alpha_2, \alpha_3)$. We may now have $V_3 = E$, but if not .... This process may or may not end in a finite number of steps. If it does, say if $E = V_F(\alpha_1, \ldots, \alpha_n)$, then $\dim_F E = n$, and by the usual theorems on the uniqueness of the dimension, we know that n is uniquely determined by E and F, independent of the choice of the α's. It is called the *degree* of E over F:

DEFINITION.   If F is subfield of E, then the *degree of E over F* (written $[E:F]$ or $[E/F]$) is defined by $[E:F] = \dim_F E$.

From the above it follows immediately that for any $F \subseteq E$ we have $[E:F] = 1 \Leftrightarrow E = F$.

A basic theorem on degrees states [A, p. 21]:

33

2.1 DEGREE OF AN ALGEBRAIC EXTENSION

THEOREM. If $F \leq K \leq E$, where F, K, E are fields, then

$$[E:F] = [E:K] \cdot [K:F].$$

Proof. Let $\alpha_1, \ldots, \alpha_r$ be basis elements of K over F, so that $[K:F] \geq r$ and let $\beta_1, \ldots, \beta_s$ be basis elements of E over K, so that $[E:K] \geq s$. We shall first show that the rs elements of the form $\alpha_i \beta_j$ $(1 \leq i \leq r, 1 \leq j \leq s)$ are linearly independent: We suppose they are not independent and shall arrive at a contradiction.

So let us suppose that for some $c_{i,j} \in F$, not all 0, we have

$$c_{1,1}\alpha_1\beta_1 + c_{2,1}\alpha_2\beta_1 + \cdots + c_{r,s}\alpha_r\beta_s = 0. \tag{1}$$

Grouping the terms containing β_j $(1 \leq j \leq s)$ we get

$$(c_{1,1}\alpha_1 + c_{2,1}\alpha_2 + \cdots + c_{r,1}\alpha_r)\beta_1 + \cdots + (c_{1,s}\alpha_1 + \cdots + c_{r,s}\alpha_r)\beta_s = 0. \tag{2}$$

Since some $c_{i,j}\alpha_1 + \cdots + c_{r,j}\alpha_r \neq 0$, this contradicts our hypothesis that the β's are linearly independent over K. Therefore, E contains at least rs elements (the $\alpha_i\beta_j$) which are linearly independent over F; that is, $[E:F] \geq rs$. If $[K:F]$ or $[E:F]$ should be infinite, this means that rs can be made arbitrarily large, so $[E:F]$ must also be infinite, and the theorem holds in this case.

If r and s are finite and $\alpha_1, \ldots, \alpha_r$ span K over F while $\beta_1, \ldots, \beta_s$ span E over K, we can show that the $\alpha_i\beta_j$ actually span E over F; that is, every element $\gamma \in E$ is a linear combination of $\alpha_i\beta_j$'s with coefficients in E: We can then write any element $\gamma \in E$ in the form

$$\gamma = b_1\beta_1 + \cdots + b_s\beta_s, \quad \text{where } b_j \in K, \quad i \leq j \leq s. \tag{3}$$

Since $b_j \in K$ and $F < K$, we can write each b_j in the form

$$b_j = c_{1j}\alpha_1 + \cdots + c_{r,j}\alpha_r$$

and substitute this into (3). So γ can be written as

$$\gamma = (c_{1,1}\alpha_1 + \cdots + c_{r,1}\alpha_1)\beta_1 + \cdots + (c_{1,s}\alpha_1 + \cdots + c_{r,s}\alpha_r)\beta_s$$

and multiplying out we get

$$\gamma = \underline{\hspace{8cm}}.$$

So E is spanned by the rs elements $\alpha_i \beta_j$, and $[E:F]$ must be exactly equal to rs. ‖

A few examples are given in Table 2.1, where $GF(p^n)$ is the Galois field containing p^n elements, $i = \sqrt{-1}$, $\omega = -\frac{1}{2} + \frac{1}{2}\sqrt{-3}$, $\pi = 3.14\ldots$.

In example 23, $Q(\pi)$ is a *transcendental* extension of Q.

DEFINITION. An element $z \in E$ is called *transcendental over F* if there is no polynomial $p(x) \in F[x]$ such that $p(z) = 0$.

For example, the constants π and e are transcendental over Q, and so is the "indeterminate" x. (The words "variable" and "indeterminate" must be used with caution.) We have

DEFINITION. An extension E of F is called *transcendental* if E contains some element that is transcendental over F. Otherwise, E is called *algebraic* over F.

EXAMPLE. In Table 2.1, example 20 gives a(n)_____ extension, whereas example 24 is a(n) _____ extension.

THEOREM. If a, b are transcendental over F, then $F(a) \cong F(b)$.

Proof. The isomorphism is given by the correspondence_____

_____. ‖

In the construction of Table 2.1 it soon became evident that it would be convenient to have a simple method for calculating the degree of an extension. For example, to calculate $[Q(\sqrt{2}, \sqrt{3}):Q]$ we might proceed as follows. Let $\alpha_1 = 1$; then $V_Q(1) = Q(\alpha_1) = Q(1) = Q$. Now let $\alpha_2 = \sqrt{2}$. Then $V_Q(1) \subseteq V_Q(1, \sqrt{2}) \subseteq Q(\sqrt{2}, \sqrt{3})$. If we let $\alpha_3 = \sqrt{3}$ [justified, since $\sqrt{3} \in (Q(\sqrt{2}, \sqrt{3}) - Q(\sqrt{2}))$], then $V_Q(1) \subseteq V_Q(1, \sqrt{2}) \subseteq V_Q(1, \sqrt{2}, \sqrt{3})$, and $\alpha_1, \alpha_2, \alpha_3$ are linearly independent. However, $V_Q(1, \sqrt{2}, \sqrt{3}) \neq Q(\sqrt{2}, \sqrt{3})$, because $\sqrt{6} \in Q(\sqrt{2}, \sqrt{3})$, whereas $\sqrt{6} \notin V_Q(1, \sqrt{2}, \sqrt{3})$, because_____

2.1 DEGREE OF AN ALGEBRAIC EXTENSION

TABLE 2.1

	E	F	$[E:F]$
1	$Q(\sqrt{5})$	Q	
2	$Q(i, \sqrt{5})$	$Q(i) = \mathscr{C}$	
3	$Q(i, i\sqrt{5})$	$\mathscr{C}$	
4	$Q(\sqrt{5}, \sqrt{-5})$	$Q(\sqrt{5})$	
5	$Q(\sqrt{2}, \sqrt{3}, \sqrt{5})$	$Q(\sqrt{2})$	
6	$Q(\sqrt{2}, \sqrt{3}, \sqrt{6})$	$Q(\sqrt{2})$	
7	$Q(\sqrt[3]{2})$	Q	
8	$Q(\omega)$	Q	
9	$Q(\sqrt{\omega})$	$Q(\omega)$	
10	$Q(i, \omega)$	$Q(\omega)$	
11	$Q(x)$	$Q(x^2)$	
12	$Q(i, x)$	$Q(x)$	
13	$Q(\sqrt{2}, \sqrt[3]{2})$	Q	
14	$Q(\sqrt{2}, \sqrt[3]{2}, \sqrt[4]{2})$	Q	
15	$Q(\sqrt[3]{2}, \sqrt[4]{2})$	Q	
16	$Q(\sqrt[3]{2}, \sqrt[4]{2}, \sqrt[5]{2})$	Q	
17	$Q(\sqrt[3]{2}, \sqrt[4]{2}, \sqrt[5]{2}, \sqrt[6]{2})$	Q	
18	$Q[x]/(x + 1)$	Q	
19	C	R	
20	R	Q	
21	$GF(2^6)$	$GF(2)$	
22	$GF(2)(x)$	$GF(2)$	
23	$Q(\pi)$	Q	
24	$Q(\sqrt{2} + \sqrt{3})$	Q	
25	$Q(\sqrt{2} \cdot \sqrt{3})$	Q	

_____. So we need at least one more basis vector, say $\alpha_4 = \sqrt{6}$, and then we must check carefully whether $V_Q(1, \sqrt{2}, \sqrt{3}, \sqrt{6}) = Q(\sqrt{2}, \sqrt{3})$ or not. In particular, it is necessary to check whether every element of the form $x_1 = (a_1 + a_2\sqrt{2} + a_3\sqrt{3} + a_4\sqrt{6}) \neq 0$ has a multiplicative inverse of the same form. We leave it as an exercise to show that this is so. (Let $x_2 = a_1 - a_2\sqrt{2} + a_3\sqrt{3} - a_4\sqrt{6}$), $x_3 = a_1 + a_2\sqrt{2} - a_3\sqrt{3} - a_4\sqrt{6}$, $x_4 = a_1 - a_2\sqrt{2} - a_3\sqrt{3} + a_4\sqrt{6}$ and show that $x_1x_2x_3x_4 \in Q$, so that

$$\frac{1}{x_1} = \frac{x_2x_3x_4}{x_1x_2x_3x_4} \in V_Q(1, \sqrt{2}, \sqrt{3}, \sqrt{6}).$$

This method of finding the degree of an extension is rather awkward even in simple situations. Another way of calculating the degree follows very naturally from the fact that $F(\alpha) \cong F[x]/(p(x))$, where $p(x)$ is a polynomial irreducible over F and with α as a root, remembering that if $\deg p(x) = n$, then the equivalence classes of $1, x, x^2, \ldots, x^{n-1}$ form a basis of $F[x]/(p(x))$, so that $[F[x]/(p(x)):F] = n$. Hence $[F(\alpha):F] = n$. Applying this to be the last situation we see easily that if $\alpha = \sqrt{2}$, then we can take $p(x) = $ _____, to get $Q(\sqrt{2}) \cong Q[x]/(p(x))$, so that $n = 2$. Then $Q(\sqrt{2}, \sqrt{3}) = (Q(\sqrt{2}))(\sqrt{3})$, and $\sqrt{3}$ is a root of $x^2 - 3$, which is irreducible over $Q(\sqrt{2})$. Hence $[Q(\sqrt{2}, \sqrt{3}):Q(\sqrt{2})] = 2$. Combining these two results, we get

$$[Q(\sqrt{2}, \sqrt{3}):Q] = [Q(\sqrt{2}, \sqrt{3}):Q(\sqrt{2})] \cdot [Q(\sqrt{2}):Q] = 2 \cdot 2 = 4,$$

which agrees with the previous answer. This second method of calculating the degree when α is an algebraic number expressed in terms of radicals is usually more practical.

In examples 20 and 23 a few more questions come up. In both cases we find that $[E:F] = \infty$. Now suppose there is a field K such that $F < K < E$ (proper inclusions). Can we find K such that $[K:F] = [E:F] = \infty$ in (20)?_____ In (23)?_____ Such that $[K:F]$ is finite in (20)?_____ In (23)?_____ Such that $[K:F]$ is infinite in (20)?_____ In (23)?_____ Clearly $[K:F]$ and $[E:F]$ cannot both

2.1 DEGREE OF AN ALGEBRAIC EXTENSION

be finite, by the theorem on page 38. Answering some of the questions above may not be easy. You might find [vdW, p. 198, Lüroth's theorem] a useful reference.

Perhaps you noticed in filling out the table of degrees that we often have $[E:F] = m \cdot n$, when $E = F(\alpha, \beta)$ or $F(\alpha + \beta)$ or $F(\alpha - \beta)$ or $F(\alpha\beta)$ or $F(\alpha^{1/n})$ with $[F(\alpha):F] = m$ and $[F(\beta):F] = n$. Is this a general theorem? Before attempting to prove such a result, let us try a few more examples, even though at first glance some of them seem trivial. Care is necessary, especially in the second set.

(1)

α	β	$[Q(\alpha, \beta):Q]$	$[Q(\alpha + \beta):Q]$	$[Q(\alpha - \beta):Q]$	$[Q(\alpha\beta):Q]$	$[Q(\alpha/\beta):Q]$
$\sqrt{2}$	$\sqrt{3}$					
$\sqrt{2}$	$\sqrt{2}$					
$\sqrt{2}$	$1 + \sqrt{2}$					
$\sqrt{2}$	$1/\sqrt{2}$					

(2)

F	E		$[E:F]$
Q	$Q(\sqrt{3})$		2
Q	$Q(\sqrt[3]{2} + \frac{10}{9}\sqrt{3})$ (real cube root)		
Q	$Q(\sqrt[3]{2} - \frac{10}{9}\sqrt{3})$ (real cube root)		

(2) (contd.)

F	E	$[E:F]$
Q	$Q(\sqrt[3]{2 + \frac{10}{9}\sqrt{3}} + \sqrt[3]{2 - \frac{10}{9}\sqrt{3}})$	1
Q	$Q(\sqrt[3]{2 + \frac{10}{9}\sqrt{3}} - \sqrt[3]{2 - \frac{10}{9}\sqrt{3}})$	
Q	$Q(\omega\sqrt[3]{2 + \frac{10}{9}\sqrt{3}} + \omega^2\sqrt[3]{2 - \frac{10}{9}\sqrt{3}})$	
Q	$Q(\omega\sqrt[3]{2 + \frac{10}{9}\sqrt{3}} - \omega^2\sqrt[3]{2 - \frac{10}{9}\sqrt{3}})$	

The last example suggests that caution is advisable in the general case, that we do not necessarily have $[E:F] = m \cdot n$, and that the answer is rarely apparent on inspection. We also see that care is necessary in the descriptions of α and β, even where they are written out in terms of radicals: In particular, whenever we take an nth root, it must be specified which of the n possible values is meant. Is it, however, true that in all cases $[E:F]$ is a divisor of mn? Can you prove it?

Suppose we know that $F(\alpha, \beta) = m \cdot n$. Does this imply that $F(\alpha + \beta) = m \cdot n$? _____ There are many more questions of this nature that one may ask, and many of them are not too difficult to answer, although they are by no means trivial. The later sections will help greatly with the answers.

2.2 ISOMORPHISMS OF FIELDS

There are many ways of defining fields: as algebraic extensions of the rationals, as the quotient of a polynomial ring by a maximal ideal $F[x]/(p(x))$, as the set of remainders

2.2 ISOMORPHISMS OF FIELDS

on division by an irreducible polynomial $p(x)$. These are not the only ways—there are fields of functions, for instance, with the field $F(x)$ of all formal ratios of polynomials in $F[x]$ (with nonzero denominator, of course) as an example. Sometimes we can show that a set of elements that remains invariant under a certain set of transformations also forms a field. Such a field is called the fixed field of this set of transformations. Galois theory has much to say about such fields.

We next give a number of examples of fields defined in various ways and bring up the question: Are these really all different fields, or are some of them perhaps essentially identical (isomorphic) but merely described in a different manner? Are some of these fields subfields or extensions of some of the others? Are some of them isomorphic, but not identical? [For example, $Q(\sqrt[3]{2}) \cong Q(\sqrt[3]{2}\omega)$.]

As a simple first example, note that

$$Q[x]/(x^2 - 2) \cong Q(\sqrt{2}),$$

because the correspondence

$$[ax + b] \xrightarrow{\varphi} \underline{\hspace{5cm}}$$

is an isomorphism. (Check this!) Is φ the only possible isomorphism?\underline{\hspace{2cm}}

In Tables 2.2 through 2.6, try to fill in any relevant empty squares with $=$, $\cong$, or $<$ if the fields named on the left and on the top are in the corresponding relation. Reading across, the fields are the same as those named on the left. Thus "$6 < 11$" means that field 6 [here $Q(\sqrt{2})$] is a subfield of field 11 [$Q(\sqrt{2}, \sqrt{3})$]. Remember that $\pi = 3.14\ldots$ and $e = 2.71\ldots$ are transcendental over Q. However, it is not known at present whether $\pi \in Q(e)$ or $e \in Q(\pi)$, or even whether $\pi + e$ is irrational, so you may have to insert question marks into a few of the empty squares.

TABLE 2.2

	Field no.:	1	2	3	4	5	6	7	8	9	10	11	12	13	14	15	16	17	18
1	Q	=																	
2	$Q(z)$		=																
3	$Q(\pi)$			=															
4	$Q(e, \pi)$				=														
5	$Q(x, y)$					=													
6	$Q(\sqrt{2})$						=				=	6 < 11							
7	$Q(i)$							=											
8	$Q(i\sqrt{2})$								=										
9	$Q(i + \sqrt{2})$									=									
10	$Q(i - \sqrt{2})$										=								
11	$Q(\sqrt{2}, \sqrt{3})$											=							
12	$Q(\sqrt{2}, -\sqrt{3})$												=						
13	$Q(\sqrt{2}, \sqrt{-3})$													=					
14	$Q(\sqrt{i})$														=				
15	$Q(\sqrt[4]{-4})$															=			
16	$Q(\sqrt{2} + \sqrt{3})$																=		
17	$Q(\sqrt{5 + \sqrt{24}})$																	=	
18	$Q[z]/(z)$	≅																	=

2.2 ISOMORPHISMS OF FIELDS

TABLE 2.3

Field no.:	1	2	3	4	5	6	7	8	9	10	11	12	13	14	15	16	17
1 Q	=																
2 $Q(i)$		=															
3 $Q[x]/(x^3 + 2)$			=														
4 $Q[x]/(x^3 - 2)$				=													
5 $Q[x]/(x^2 + 1)$					=												
6 $Q[x]/(x + 1)$						=											
7 $Q(\sqrt{2})$							=										
8 $Q(\sqrt{-2})$								=									
9 $Q(\sqrt[3]{2})$									=								
10 $Q(\sqrt[3]{-2})$										=							
11 $Q(\sqrt[3]{2} \cdot \omega)$											=						
12 $Q(\sqrt[3]{2} \cdot \omega^2)$												=					
13 $Q(\sqrt[3]{2} + \omega)$													=				
14 $Q[x]/(x^2 + 2)$														=			
15 $Q[x]/(x^2 - 2)$															=		
16 $Q[x]/(x - 2)$																=	
17 $Q[x]/(x + 2)$																	=

TABLE 2.4

	Field no.:	1	2	3	4	5	6	7	8	9
1	$Q(\omega)$	=								
2	$Q(\sqrt{\omega})$		=							
3	$Q(\omega^2)$			=						
4	$Q(\pi)$				=					
5	$Q(z)$					=				
6	$Q(z^2)$						=			
7	$Q(z^3)$							=		
8	$Q[z]/(z^2 + z + 1)$								=	
9	$F = \{q(z) \in Q(z) \mid q(z) = q(-z)\}$									=

2.2 ISOMORPHISMS OF FIELDS

TABLE 2.5

		Field no.:	1	2	3	4
	1	$Q(\sqrt{2})$				
	2	$Q(i\sqrt{2})$				
	3	$Q(2^{3/4})$				
	4	$Q(i2^{1/4})$				

TABLE 2.6

		Field no.:	1	2	3	4	5	6	7	8
	1	GF(2)	=							=
	2	GF(2^2)		=						
	3	$Z/(2)$			=					
	4	GF(2)$[x]/(x)$				=				
	5	GF(2)$[x]/(x^2 + x + 1)$					=			
	6	GF(2)$[x]/(x^3 + x + 1)$						=		
	7	GF(2)$[x]/(x^3 + x^2 + 1)$							=	
	8	$Z[i]/(1 + i)$								=

TABLE 2.7

	Field no.:	1	2	3	4	5	6	7	8	9
1	R									
2	R(x)/(x + 1)									
3	R(i)									
4	R[x]/(x² × 1)									
5	R[x]/(x² + 2)									
6	R[x, y]/I (see note)									
7	Q(i)									
8	Q[x]/(x² + 1)									
9	Q[x]/(x² + 2)									

Note: In (6) the ideal I is the set of all polynomials of R[x, y] that have no constant term. Is I maximal in R[x, y]? _____ . Is R[x, y]/I really a field? _____ .

2.2 ISOMORPHISMS OF FIELDS

EXAMPLE.   To show that $Q(\sqrt{i}) = Q(i, \sqrt{2})$.

From DeMoivre's theorem we have

$$\sqrt{i} = \left(\cos\frac{\pi}{2} + i\sin\frac{\pi}{2}\right)^{1/2} = \cos\frac{\pi}{4} + i\sin\frac{\pi}{4} = \frac{1}{\sqrt{2}} + i\frac{1}{\sqrt{2}}.$$

Also,

$$-i\sqrt{i} = \underline{\hspace{3cm}}.$$

Subtracting, we get

$$\sqrt{i} - i\sqrt{i} = \underline{\hspace{3cm}}.$$

Also

$$i = (\sqrt{i})^2.$$

This shows that $Q(\sqrt{2}, i) \subseteq Q(\sqrt{i})$. Taking this together with the expression $\sqrt{i} = (1/\sqrt{2}) + i(1/\sqrt{2})$, we see also that $Q(\sqrt{i}) \subseteq Q(\sqrt{2}, i)$; hence the fields are the same.

EXERCISE.   Show that $Q(\sqrt{2}) \ncong Q(\sqrt{3})$.

Filling in Tables 2.2 to 2.7 naturally brings up the question as to whether there is an easy way to do this. There is not. But is there any systematic method to determine whether $F_1 < F_2$ whenever F_1, F_2 are two given fields? The answer to this depends entirely on what is meant by the word "given." Obviously, if say $F_1 = Q[x]/(p_1(x))$, $F_2 = Q[x]/(p_2(x))$, where $p_1 \neq p_2 \in Q[x]$ and irreducible and their coefficients are explicitly stated [say, for example, $p_1(x) = x^5 + x + 1$], then $F_1 \neq F_2$, and the only sensible question one can ask is whether $F_1 \cong F_2$. In this case there is a systematic procedure, but it is fairly complicated and cannot be considered an easy way out. (We shall return to this matter later.)

When studying groups, rings, and other algebraic structures one hears a great deal about their homomorphisms and homomorphic images. With fields the story is, however, different: We always consider isomorphisms. The explanation is very simple:

A homomorphism φ is an isomorphism if and only if $\varphi(a) = 0$ implies that $a = 0$. So suppose F is a field and φ a homomorphism but not an isomorphism. Then

$$(\exists a)(a \neq 0 \text{ and } \varphi(a) = 0).$$

Now let b be any other element of F. Since F is a field,

$$(\exists c)(b = ac, c \in F).$$

Then $\varphi(b) = $ _____, so the image of F consists only of one element, $\varphi(F) = \{\varphi(b)|b \in F\} = \{0\}$, and so φ is the trivial homomorphism.

Although it is not relevant to the study of Galois theory, it is perhaps of interest to make a few remarks about *ordered* fields and their isomorphisms:

DEFINITION. A field F is an *ordered* field if there is a subset $P \subseteq F$ such that

(1) For every $a \in F$ exactly one of the following holds:

$$a \in P \qquad -a \in P \qquad \text{or} \qquad a = 0.$$

(2) If $a, b \in P$, then $a + b \in P$ and $ab \in P$.

If we write $a < b$ whenever $(b - a) \in P$, then any set P satisfying (1) and (2) completely determines the ordering "$<$" of F, because _____

_____.

DEFINITION. The field A of all *algebraic numbers* is the set of all roots of some polynomial over Q,

$$A = \{\alpha | p(\alpha) = 0 \text{ for some } p(x) \in Q[x]\}$$

and the field A_r of all real *algebraic* numbers is

$$A_r = A \cap \mathrm{R}.$$

2.2   ISOMORPHISMS OF FIELDS

THEOREM.   Each of the three fields Q, R, and A_r can be ordered in one and only one way.

Proof. For Q: Either $1 \in P$ or $-1 \in P$, and not both. But $(-1) \in P \Rightarrow (-1)(-1) = 1 \in P$, which is impossible, so we must have $1 \in P$, and also $\underbrace{1 + \cdots + 1}_{n\,\text{times}} = n \in P$.

Suppose $n \in P$ and $1/n \notin P$. Then $-1/n \in P$ and $(n)(-1/n) = -1 \in P$, again a contradiction. Therefore, if $m = \underbrace{1 + \cdots + 1}_{m\,\text{times}}$, then $m/n \in P$, which gives the usual ordering of Q.

For R and A_r: Both fields have the property that for every element α either α or $-\alpha$ has a square root that is also an element of the field. Now if $\beta = \sqrt{\alpha}$, then $\beta \in P$ or $-\beta \in P$. In either case $\alpha = \beta \cdot \beta = (-\beta)(-\beta) \in P$, so P contains all those elements which are squares. It cannot contain any others; for suppose $\alpha \in P$ and has no square root. Then $(-\alpha)$ does have a square root, say $\sqrt{-\alpha} = \gamma$, so that $(-\alpha) = \gamma^2 \in P$. But then $\alpha \notin P$, so we are led to a contradiction. Therefore, in both fields the set P of positive elements coincides with the set of all elements that have square roots in the field. Since this set is uniquely determined, so is P.   ‖

On the other hand, we have

THEOREM.   The fields $\mathscr{C}$, C, and A cannot be ordered.

Proofs.   In each of these fields we have an element $i = \sqrt{-1}$. If the field could be ordered, then $i \in P$ or $-i \in P$, so suppose that $+i \in P$ or $-i \in P$. Then $(\pm i)(\pm i) = -1 \in P$ and $1 \notin P$. This is impossible, for $1 \in P$ because _____

_____. Therefore, neither $i \in P$ nor $-i \in P$, and so the fields cannot be ordered.   ‖

We ask, therefore:

(1) Can any real field F with $F \subseteq R$ be ordered in *more* than one way?

(2) Is there any field F that is not a real field (that is, F does contain some complex numbers) and that can be ordered?

As usual, it is wise to look first at specific numerical examples before trying to give a general answer.

EXAMPLES.   (1) Let $F_1 = Q(\sqrt{2})$,

$$\varphi(a + b\sqrt{2}) = a - b\sqrt{2}, \qquad a, b \in Q.$$

Then φ maps F_1 isomorphically onto itself. Now let

$$P_1 = \{a + b\sqrt{2}|(a + b\sqrt{2}) > 0\},$$

$$P_2 = \varphi(P_1) = \{y|(\exists x)(x \in P_1 \text{ and } y = \varphi(x)\}$$

$$= \{\varphi(a + b\sqrt{2})|(a + b\sqrt{2}) > 0\}$$

$$= \{(a - b\sqrt{2})|(a + b\sqrt{2}) > 0\}.$$

Clearly $P_1 \neq P_2$, because _____. However,

$$P_1\begin{cases} \text{and } P_2 \\ \\ \text{but not } P_2 \end{cases}$$ (cross out one) will satisfy the conditions for a set P of positive elements

of an ordered field. So the answer to question (1) is _____.

   (2) Let

$$F_2 = Q(\sqrt[4]{2}),$$

$$F_3 = Q(i\sqrt[4]{2}),$$

$$\varphi(\sqrt[4]{2}) = i\sqrt[4]{2}.$$

Clearly F_2 is a real field that can be ordered in a natural way and φ determines an isomorphism of F_2 onto F_3.

Now let

$$P_2 = \{x|x \in F_2 \text{ and } x > 0\},$$

$$P_3 = \varphi(P_2).$$

Since F_2 is ordered, P_2 is the natural set of positive elements, but one must check that

P_3 is acceptable as a set of positive elements of F_3. This is easily done by checking the two conditions P_3 must satisfy:

(1) Suppose

$$0 \neq a = c_0 + c_1(i2^{1/4}) + c_2(-2^{1/2}) + c_3(-i2^{3/4})$$

is any element of F_3. Then

$$a = \varphi(c_0 + c_1 2^{1/4} + c_2 2^{1/2} + c_3 2^{3/4}) = \varphi(a'),$$

where $a' \in F_2$. We know that $a' \in P_2$ or $-a' \in P_2$, so we know that $a \in P_3$ or $-a \in P_3$.

(2) Suppose

$$a = \varphi(a'), b = \varphi(b'), \qquad \text{with } a, b \in P_3.$$

Then $a', b' \in P_2$, and since F_2 is ordered, so are $a' + b'$ and $a'b'$. Therefore, $a + b = \varphi(a') + \varphi(b') = \varphi(a' + b') \in P_3$ and $ab = \varphi(a')\varphi(b') = \varphi(a'b') \in P_3$. Therefore, F_3 can be ordered, even though it is not a real field. Can F_2 be ordered in more than one way?

We shall close this section with some more questions whose answers are left as exercises:

(1) Are there real fields that can be ordered in three different ways? In n different ways? Can you construct examples?

(2) Is the following conjecture true or false: A field $F \subset A_r$ can be ordered if and only if it is isomorphic to some real field?

2.3 AUTOMORPHISMS OF FIELDS

DEFINITION. A mapping φ is called an *automorphism* of the field F if it is an isomorphism of F onto itself.

Notice that from Table 2.4 we see that it is possible to map a field F isomorphically into (and not onto!) itself, for $Q(z) \cong Q(z^2)$ and $Q(z^2) \subsetneqq Q(z)$, but such a mapping is not called an automorphism.

A field F can often be mapped isomorphically onto itself in several different ways. A simple example is the following:

EXAMPLES. (1) Let $F = Q(\sqrt[3]{2}, \omega)$.

As first automorphism φ_1 we simply take the identity mapping of F onto itself. Another automorphism φ_2 of F is uniquely determined if we specify

(1) $r \in Q \Rightarrow \varphi_2(r) = r$,

(2) $\varphi_2(\sqrt[3]{2}) = \omega\sqrt[3]{2}$,

(3) $\varphi_2(\omega) = \omega^2$,

because every element of F is a polynomial in $\sqrt[3]{2}$ and ω.

For example,

$$\varphi_2(\tfrac{5}{6} + 6\sqrt[3]{2} - \omega\sqrt[3]{2} + 2^{2/3}) = \underline{\hspace{4cm}}.$$

It is, of course, necessary to verify that the correspondence determined in this seemingly arbitrary manner is indeed a homomorphism: This can be checked rather routinely:

Every element α of $Q(\sqrt[3]{2}, \omega)$ can be written in the form

$$\alpha = a + b2^{1/3} + c2^{2/3} + d\omega + e\omega2^{1/3} + f\omega2^{2/3}.$$

(Why don't we need any terms involving ω^2 in this expression?)

If we let

$$\alpha' = a' + b'\sqrt{2} + \underline{\hspace{5cm}} + f'\omega2^{2/3},$$

we see that

$$(\alpha + \alpha') = \underline{\hspace{6cm}},$$

$$\varphi_2(\alpha + \alpha') = \underline{\hspace{5cm}},$$

2.3 AUTOMORPHISMS OF FIELDS

whereas

$$\varphi_2(\alpha) = \underline{\hspace{8cm}},$$

$$\varphi_2(\alpha') = \underline{\hspace{8cm}},$$

so that

$$\varphi_2(\alpha) + \varphi_2(\alpha') = \underline{\hspace{6cm}} = \varphi_2(\alpha + \alpha'),$$

which checks the addition property.

To check that φ_2 also commutes with multiplication, that is, $\varphi_2(\alpha \cdot \alpha') = \varphi_2(\alpha) \cdot \varphi_2(\alpha')$, just replace addition by multiplication in this calculation.

In just the same way, one can check that each of the following mappings is also an automorphism of F:

φ_3 defined by (1) $r \in Q \Rightarrow \varphi_2(r) = r$,

                    (2) $\varphi_3(2^{1/3}) = 2^{1/3}$,

                    (3) $\varphi_3(\omega) = \omega^2$

φ_4 defined by (1) $r \in Q \Rightarrow \varphi_3(r) = r$,

                    (2) $\varphi_4(2^{1/3}) = 2^{1/3}\omega$,

                    (3) $\varphi_4(\omega) = \omega$,

φ_5 defined by (1) $r \in Q \Rightarrow \varphi_4(r) = r$,

                    (2) $\varphi_5(2^{1/3}) = 2^{1/3} \cdot \omega^2$,

                    (3) $\varphi_5(\omega) = \omega$,

φ_6 defined by (1) $r \in Q \Rightarrow \varphi_6(r) = r$,

                    (2) $\varphi_6(2^{1/3}) = 2^{1/3}\omega^2$,

                    (3) $\varphi_6(\omega) = \omega^2$.

We have here listed six different automorphisms of F. Are there any others?

(2) In Table 2.8 we list more mappings of fields into themselves. Some of these mappings are automorphisms, some are not.

TABLE 2.8

	E	φ	Is φ an automorphism?	Is φ an isomorphism?
1	Q	$r \to r + 1$		
2	$Q(\sqrt{2})$	$\sqrt{2} \to -\sqrt{2}$		
3	$Q(\sqrt[3]{2})$	$\sqrt[3]{2} \to -\sqrt[3]{2}$		
4	$Q(\sqrt{2}, \sqrt{-2})$	$\sqrt{2} \to \sqrt{-2}$ $\sqrt{-2} \to \sqrt{2}$		
5	$Q(i)$	$i \to -i$		
6	$Q(\omega)$	$\omega \to -\omega$		
7	$Q(\sqrt[3]{5}, \omega)$	$\sqrt[3]{5} \to \omega\sqrt[3]{5}$ $\omega \to 1$		
8	$Q(z)$	$z \to z + 1$		
9	$Q(z)$	$z \to z^3$		
10	$Q(z)$	$z \to \dfrac{1}{z}$		

EXERCISES. (1) Show that if φ is an automorphism of Q, then φ is the identity mapping.

(2) The mapping $\varphi : \sqrt{2} \to -\sqrt{2}$ of $Q(\sqrt{2})$ is an automorphism, but it is not order-preserving.

2.3 AUTOMORPHISMS OF FIELDS

(3) In Table 2.9 list all possible automorphisms of $E = Q(\sqrt[3]{2}, \sqrt[3]{5}, \omega)$. There are 18 of them. From this table, we see that

$$\{\varphi | \varphi(\sqrt[3]{2}) = \sqrt[3]{2}\} = \{\varphi_1, \varphi_2, \underline{\hspace{3cm}}\},$$

$$\{\varphi | \varphi(\sqrt[3]{5}) = \sqrt[3]{5}\} = \{\varphi_1, \varphi_2, \underline{\hspace{3cm}}\},$$

$$\{\varphi | \varphi(\omega) = \omega\} = \{\varphi_1, \underline{\hspace{3cm}}\},$$

$$\{\varphi | \varphi(\omega\sqrt[3]{2}) = \omega\sqrt[3]{2}\} = \{\varphi_1, \underline{\hspace{3cm}}\},$$

$$\{\varphi | \varphi(\sqrt[3]{2}) = \sqrt[3]{2} \text{ and } \varphi(\omega) = \omega\} = \{\varphi_1, \underline{\hspace{3cm}}\}.$$

We shall return to this example in Section 3.2.

In all our examples we have

$$r \in Q \Rightarrow \varphi(r) = r.$$

Could we, with a little more imagination, have produced an automorphism in which some rational was not mapped onto itself? We note the following: For any automorphism φ of a field F,

(1) $\varphi(0) = 0$, because $\underline{\hspace{6cm}}$.

(2) $\varphi(1) = 1$, because $\underline{\hspace{6cm}}$.

(3) $\varphi(\underbrace{1 + 1 + \cdots + 1}_{m \text{ times}}) = \underbrace{1 + \cdots + 1}_{m \text{ times}}$, because $\underline{\hspace{4cm}}$.

(4) $\varphi\left(\dfrac{m}{n}\right) = \dfrac{m}{n}$, because $\underline{\hspace{6cm}}$.

(5) So if $Q < F$, then φ is the identity mapping on Q, by (1) through (4).

(6) If $\alpha \in F$ has the property that $\alpha^3 = r \in Q$, then

$$(\varphi(\alpha))^3 = \varphi(\alpha)\varphi(\alpha)\varphi(\alpha) = \varphi(\alpha^3) = \varphi(r) = r,$$

and, more generally, if $p(x) \in Q[x]$ and $p(\alpha) = 0$, then $\varphi(p(\alpha)) = p(\varphi(\alpha)) = 0$, because

TABLE 2.9

φ	$\sqrt[3]{2} \rightarrow$	$\sqrt[3]{5} \rightarrow$	$\omega \rightarrow$
φ_1	$\sqrt[3]{2}$	$\sqrt[3]{5}$	ω
φ_2	$\sqrt[3]{2}$	$\sqrt[3]{5}$	ω^2
φ_3	$\sqrt[3]{2}$	$\omega\sqrt[3]{5}$	ω
φ_{18}			

2.3 AUTOMORPHISMS OF FIELDS

(7) No automorphism φ of $Q(\sqrt[3]{2}, \omega)$ therefore could have interchanged, say ω and $2^{1/3}$, for ω is a root of the irreducible polynomial _____ $\in Q[x]$ and $\sqrt[3]{2}$ is a root of _____.

(8) So if $\alpha_1, \ldots, \alpha_n \in F$, an automorphism φ of the field $F \supseteq Q$, being one to one, will permute the roots $\alpha_1, \ldots, \alpha_n$ of any irreducible polynomial in $Q[x]$. (If φ leaves the roots fixed, it counts as the identity permutation.)

(9) The next question is evidently this: Suppose we are given an irreducible polynomial $p(x) \in Q[x]$ with roots $\alpha_1, \ldots, \alpha_n$ in F. Will every permutation π of these roots give us an automorphism of F? In other words, given any permutation π of the roots, is there always an automorphism φ such that

$$\varphi(\alpha_i) = \pi(\alpha_i), \quad \text{for } i = 1, \ldots, n?$$

(10) To answer the last question, suppose we consider the equation $p(x) = x^3 - 1 = 0$, and let $F = Q(i\sqrt{3})$. The roots are

$$\alpha_1 = 1,$$

$$\alpha_2 = \omega,$$

$$\alpha_3 = \omega^2.$$

Suppose $\pi = (123)$; is there an automorphism φ such that

$$\varphi(\alpha_1) = \alpha_{\pi(1)} = \quad \alpha_2 \quad = \underline{\hspace{3cm}},$$

$$\varphi(\alpha_2) = \alpha_{\pi(2)} = \underline{\hspace{2cm}} = \underline{\hspace{3cm}},$$

$$\varphi(\alpha_3) = \alpha_{\pi(3)} = \underline{\hspace{2cm}} = \underline{\hspace{3cm}}?$$

Since $\alpha_1 \in Q$, we must have $\varphi(\alpha_1) \in Q$ by (5), so $\varphi(\alpha_1)$ cannot be α_2, and thus the answer to the last question is necessarily no. Is this due to the fact that $(x^3 - 1)$ can be factored? Suppose we consider only irreducible polynomials $p(x)$, say

$$p(x) = x^4 + x^3 + x^2 + 1,$$

would we then be able to find an automorphism of $Q(\alpha_1, \ldots, \alpha_4)$ where $\alpha_1, \alpha_2, \alpha_3, \alpha_4$ are the roots of this polynomial, for each permutation of the roots? To answer, note that $\alpha_1^5 = \alpha_2^5 = \alpha_3^5 = \alpha_4^5 = 1$, because _____,

so if we let ζ be a primitive fifth root of unity, that is, $1 \neq \zeta = \sqrt[5]{1}$, then the roots are (suitably renumbered)

$$\alpha_1 = \zeta,$$

$$\alpha_2 = \zeta^2,$$

$$\alpha_3 = \zeta^3,$$

$$\alpha_4 = \zeta^4.$$

Again let φ correspond to (123), that is, $\varphi(\alpha_1) = \alpha_2 = $ _____, $\varphi(\alpha_2) = \alpha_3 = $ _____, and $\varphi(\alpha_3) = $ _____. But also

$$\varphi(\alpha_2) = \varphi(\zeta^2) = \varphi(\zeta)\varphi(\zeta) = \varphi(\alpha_1)\varphi(\alpha_1) = \underline{\hspace{2cm}} \cdot \underline{\hspace{2cm}} = \zeta^4.$$

So this correspondence does not have the multiplicative property of automorphisms and therefore cannot be an automorphism.

(11) Our conclusion therefore is: If F contains all the roots of the polynomial $p(x)$ and φ is an automorphism of F, then φ will effect a certain permutation of the roots of $p(x)$, but in general not all permutations are possible.

2.4 FIXED FIELDS

Let E be a field, φ an automorphism of E, and $F_\varphi = \{x \mid x \in E$ and $\varphi(x) = x\}$; that is, F_φ is the set of those elements of E that are left fixed by φ. Then F_φ is a field, because _____.

We therefore make the following

2.4 FIXED FIELDS

DEFINITION. If φ is an automorphism of the field E, then $F_\varphi = \{x | x \in E$ and $\varphi(x) = x\}$ is called the *fixed field of E under* φ.

DEFINITION. If S is a set of automorphisms of E, then $F_S = \{x | x \in E$ and $\forall \varphi(\varphi \in S \Rightarrow \varphi(x) = x)\}$ is called the *fixed field of E under S.*

   In this book, the set S will usually be a group.

EXAMPLES. In the next set of tables, φ is described by specifying its effect on the generating elements of the field; F_φ, F_S, and the degrees $[E:F_\varphi]$, $[E:F_S]$ are to be filled in. For instance, $\sqrt{2} \to -\sqrt{2}$, $i \to -i$ means that $\varphi(\sqrt{2}) = -\sqrt{2}$, $\varphi(i) = -i$. As we saw, every rational number is necessarily left invariant. In the last column, left blank, make up an example of your own, perhaps somewhat more complicated than the preceding.

E	$Q(i)$	$Q(\omega)$	$Q(z)$	$Q(z)$	$Q(x_1, x_2, \ldots, x_n)$
φ	$i \to -i$	$\omega \to \omega^2$	$z \to \dfrac{1}{z}$	$z \to z^2$	$x_1, x_2, \ldots, x_n \to$ any permutation of $x_1, x_2, \ldots, x_n$
F_φ					
$[E:F_\varphi]$					See [A, p. 39]

Also $a \to a^k$ means that $\varphi(a) = a^k$ whenever a is any element of the field. Always check carefully whether φ is really a homomorphism.

E	$Q(\sqrt{2},\sqrt{3},\sqrt{5})$	$Q(\sqrt{2},\sqrt{-2})$	$Q(\sqrt{2},\sqrt[3]{2})$	
φ	$\begin{cases} \sqrt{2} \to -\sqrt{2} \\ \sqrt{3} \to \sqrt{3} \\ \sqrt{5} \to \sqrt{5} \end{cases}$	$\begin{cases} \sqrt{2} \to \sqrt{2} \\ \sqrt{-2} \to -\sqrt{2} \end{cases}$	$\begin{cases} \sqrt{2} \to -\sqrt{2} \\ \sqrt[3]{2} \to -\sqrt[3]{2} \end{cases}$	
F_φ			None; φ is not a homomorphism. (Why?)	
$[E:F_\varphi]$				

E	$GF(2^n)$	$GF(2^n)$	$GF(3^n)$	$GF(3^n)$
φ	$a \to a^2$	$a \to a^3$	$a \to a^2$	$a \to a^3$
F_φ				
$[E:F_\varphi]$				

In the next table we let φ_1 be the identity mapping in each case and let

$S_1 = \{\varphi_1, \varphi_2\}$, where $\varphi_2 : i \to -i$,

$S_2 = \{\varphi_1, \varphi_2\}$, where $\varphi_2 : \begin{cases} \sqrt{2} \to -\sqrt{2} \\ \sqrt[3]{2} \to \sqrt[3]{2}, \end{cases}$

$S_3 = \{\varphi_1, \varphi_2\}$, where $\varphi_2 : \begin{cases} \sqrt[3]{2} \to \sqrt[3]{2}\omega \\ \omega \to \omega^2, \end{cases}$

$S_4 = \{\varphi_1, \varphi_3\}$, where $\varphi_3 : \begin{cases} \sqrt[3]{2} \to \sqrt[3]{2} \\ \omega \to \omega^2, \end{cases}$

$S_5 = \{\varphi_1, \varphi_4, \varphi_5\}$, where $\varphi_4 : \begin{cases} \sqrt[3]{2} \to \sqrt[3]{2}\omega \\ \omega \to \omega \end{cases}$ and $\varphi_5 : \begin{cases} \sqrt[3]{2} \to \sqrt[3]{2}\omega^2 \\ \omega \to \omega. \end{cases}$

2.4 FIXED FIELDS

E	$Q(i)$	$Q(\sqrt{2}, \sqrt[3]{2})$	$Q(\sqrt[3]{2}, \omega)$	$Q(\sqrt[3]{2}, \omega)$	$Q(\sqrt[3]{2}, \omega)$
S	S_1	S_2	S_3	S_4	S_5
F_S					
$[E:F_S]$					

Next we look at the problem in reverse: Given a field E and a subfield F, is it always possible to find a set S of automorphisms of E such that the fixed field F_S of S is precisely equal to F? If we can, then S would characterize the field F. Let us try an example. Let $E = Q(\sqrt[3]{2}, i)$ and $F = Q$. Any automorphism φ of E must send $\sqrt[3]{2}$ into another cube root of 2 and i into $\pm i$. Since E contains only one cube root of 2, any automorphism must therefore leave $\sqrt[3]{2}$ fixed, but we can have $\varphi(i) = i$ or $-i$. Therefore, if $\varphi(i) = i$, then $F_\varphi = E$, and if $\varphi(i) = -i$, then $F_\varphi = Q(\sqrt[3]{2})$. No other automorphisms are possible, so we can never have $F = F_S$. Clearly we were asking too much when we wanted to characterize F by finding a set S for which $F_S = F$. You should look back at this example and also at Table 2.10 when reading the section on normal extensions.

In Table 2.10 insert only those φ for which $F \leq F_\varphi$. Here $\varepsilon_8 = (1/\sqrt{2})(1 + i)$ is a primitive eighth root of unity. If φ is an automorphism of $Q(\varepsilon_8)$, it must send ε_8 into another primitive eighth root of unity, because _____

_____. The possibilities are $\varphi(\varepsilon_8) = \varepsilon_8$ or ε_8^3 or _____ or _____. In this table, z is assumed to be transcendental over Q. Apparently sometimes $F \not\leqq F_\varphi$ and sometimes $F = F_\varphi$.

If we now look at the set S of *all* automorphisms which leave the subfield F of E fixed elementwise, we notice that in each case

(1) S is a group with composition of automorphisms as the group operation. ($\varphi_1 * \varphi_2$ is φ_1 followed by φ_2.)

TABLE 2.10

If E is	and F is	then F is fixed under each of the following automorphisms φ of E	and φ leaves fixed each element of the following field F_φ:	Therefore, the field F_S left fixed by all these φ's is	The $[E:F]$ is	and $[E:F_S]$ is
$Q(\varepsilon_8)$	$Q(i)$; notice that $\varepsilon_8^2 = i$!	$\varphi_1 : \varepsilon_8 \to \varepsilon_8$	E	$Q(i)$	2	2
		$\varphi_2 : \varepsilon_8 \to \varepsilon_8^5$	$Q(i)$			
$Q(\varepsilon_8, \sqrt[3]{2})$	Q	$\varphi_1 : \varepsilon_8 \to \varepsilon_8, \sqrt[3]{2} \to \sqrt[3]{2}$	E	$Q(\sqrt[3]{2})$		
		$\varphi_2 : \varepsilon_8 \to \varepsilon_8^3, \sqrt[3]{2} \to \sqrt[3]{2}$	$Q(\sqrt[3]{2})$			
		$\varphi_3 : \varepsilon_8 \to \varepsilon_8^5, \sqrt[3]{2} \to \sqrt[3]{2}$	$Q(\sqrt[3]{2}, i)$			
		$\varphi_4 : \varepsilon_8 \to \varepsilon_8^7, \sqrt[3]{2} \to \sqrt[3]{2}$	$Q(\sqrt[3]{2})$			
$Q(\varepsilon_8, \omega)$	$Q(\varepsilon_8)$	$\varphi_1 : \omega \to \omega$				
		$\varphi_2 : \omega \to \omega^2$				
$Q(z)$	$Q(z^2)$	$\varphi_1 : z \to z$	E			
		$\varphi_2 : z \to -z$	$Q(z^2)$			
$Q(z)$	$Q(z^3)$	$\varphi_1 : z \to z$	E			

2.4   FIXED FIELDS

TABLE 2.10 (contd.)

If E is	and F is	then F is fixed under each of the following automorphisms φ of E	and φ leaves fixed each element of the following field F_φ:	Therefore, the field F_S left fixed by all these φ's is	The $[E:F]$ is	and $[E:F_S]$ is
$Q(z, i)$	$Q(z)$	$\varphi_1: i \to i$				
		$\varphi_2: i \to -i$				
$Q(z, i)$	$Q(i)$	$\varphi_1: z \to z$				
		$\varphi_2: z \to -z$				
		$\varphi_3: z \to iz$				
		$\varphi_4: z \to -iz$				
$Q(z, i)$	$Q(z^2)$	$\varphi_1: z \to z, i \to i$				
		$\varphi_2: z \to z, i \to -i$				
		$\varphi_3: z \to -z, i \to i$				
		$\varphi_4: z \to -z, i \to -i$				

(2) If $F_{S'}$ is the field left fixed by all the automorphisms of some subset $S' \subseteq S$ containing n automorphisms of E, then $[E:F_{S'}] \geq n$. Here S' need not even be a group [A, p. 36].

Do these properties hold in general? To show this, we first point out that the proof of (1) is quite easy and straightforward and is also left as an exercise. We next prove (2):

THEOREM. If $S' = \{\varphi_1, \ldots, \varphi_n\}$ is a set of n different automorphisms of E and $F_{S'}$ is the fixed field of S', then $[E : F_{S'}] \geq n$.

Proof. Suppose instead that $[E:F_{S'}] = r < n$ and that $\alpha_1, \ldots, \alpha_r$ are a basis of E over $F_{S'}$ so $E = F_{S'}(\alpha_1, \ldots, \alpha_r)$. Now consider the set of r linear homogeneous equations

$$\left.\begin{aligned}
\sum_i \varphi_i(\alpha_1)x_i &= \varphi_1(\alpha_1)x_1 + \varphi_2(\alpha_1)x_2 + \cdots + \varphi_n(\alpha_1)x_n = 0 \\
\sum_i \varphi_i(\alpha_2)x_i &= \varphi_1(\alpha_2)x_1 + \varphi_2(\alpha_2)x_2 + \cdots + \varphi_n(\alpha_2)x_n = 0 \\
&\ \ \vdots \\
\sum_i \varphi_i(\alpha_r)x_i &= \varphi_1(\alpha_r)x_1 \quad \varphi_2(\alpha_r)x_2 + \cdots + \varphi_n(\alpha_r)x_n = 0
\end{aligned}\right\} \qquad (1)$$

in $x_1, \ldots, x_n$. Since $n > r$, there are more unknowns than equations, and so we know there are $x_1, \ldots, x_n \in E$ which will satisfy these equations and are not all 0. Now let

$$\alpha = a_1\alpha_1 + a_2\alpha_2 + \cdots + a_r\alpha_r, \qquad \text{where } a_i \in F_{S'}$$

by any element of E. For any automorphism φ in S' we get

$$\begin{aligned}
\varphi(\alpha) &= \varphi(a_1)\varphi(\alpha_1) + \varphi(a_2)\varphi(\alpha_2) + \cdots + \varphi(a_r)\varphi(\alpha_r) \\
&= a_1\varphi(\alpha_1) + a_2\varphi(\alpha_2) + \cdots + a_r\varphi(\alpha_r).
\end{aligned} \qquad (2)$$

2.4 FIXED FIELDS

Multiplying the first of equations (1) by a_1, the second by a_2, and so on, and adding gives

$$a_1\varphi_1(\alpha_1)x_1 + a_1\varphi_2(\alpha_1)x_2 + \cdots + a_1\varphi_n(\alpha_1)x_n$$

$$+ a_2\varphi_1(\alpha_2)x_1 + a_2\varphi_2(\alpha_2)x_2 + \cdots + a_2\varphi_n(\alpha_2)x_n$$

$$+ \cdots$$

$$+ a_r\varphi_1(\alpha_r)x_1 + a_r\varphi_2(\alpha_r)x_2 + \cdots + a_r\varphi_n(\alpha_r)x_n = 0.$$

Carrying out the addition by columns and using (2) gives

$$x_1\varphi_1(\alpha) + x_2\varphi_2(\alpha) + \cdots + x_n\varphi_n(\alpha) = 0, \qquad \text{for every } \alpha \in E. \tag{3}$$

We show next that such a relation among n different automorphisms of E would lead to a contradiction.

Among all the solution sets for $x_1, \ldots, x_n$ there will be some which contain a maximal number of x_i's equal to zero. Suppose we have such a set and that $x_{i_1}, \ldots, x_{i_k}$ are all $\neq 0$ while all other x's vanish. Then

$$x_{i_1}\varphi_{i_1}(\alpha) + \cdots + x_{i_k}\varphi_{i_k}(\alpha) = 0, \tag{4}$$

and there is no shorter relation of this form. We cannot have $k = 1$, for then we would have $\varphi_{i_1}(\alpha) = 0$ for all α and this is impossible. Therefore, $\varphi_{i_1} \neq \varphi_{i_k}$; that is, there is some $\beta \in E$ for which $\varphi_{i_1}(\beta) \neq \varphi_{i_k}(\beta)$. Since equation (3) holds for all $\alpha \in E$ it also holds for $\alpha\beta$, so we have

$$x_{i_1}\varphi_{i_1}(\alpha\beta) + \cdots + x_{i_k}\varphi_{i_k}(\alpha\beta)$$

$$= x_{i_1}\varphi_{i_1}(\alpha)\varphi_{i_1}(\beta) + \cdots + x_{i_k}\varphi_{i_k}(\alpha)\varphi_{i_k}(\beta) = 0. \tag{5}$$

If we multiply equation (4) by $\varphi_{i_k}(\beta)$ and subtract equation (5) from it, we get

$$x_{i_1}(\varphi_{i_1}(\beta) - \varphi_{i_k}(\beta))\varphi_{i_1}(\alpha) + \cdots + x_{i_k}(\varphi_{i_k}(\beta) - \varphi_{i_k}(\beta))\varphi_{i_k}(\alpha) = 0. \tag{6}$$

The coefficient of $\varphi_{i_1}(\alpha)$ in equation (6) is not 0, because _____
_____, whereas the coefficient of $\varphi_{i_k}(\alpha)$ is 0. Equation (6) is therefore a

relationship in which at most $k - 1$ of the coefficients are different from 0. This contradicts the hypothesis, so no relation of the form (3) is possible.

We therefore cannot have $[E:F_{s'}] = r < n$, which proves the theorem. ||

Some very interesting applications of this theorem will be found in [A, p. 38]. The set S' of this theorem need not be a group, but in the important special case that it is, we can sharpen the result and obtain equality:

THEOREM.   If $G = \{\varphi_1, \ldots, \varphi_n\}$ is a group of n different automorphisms of E and F_G is the fixed field of G, then $[E:F_G] = n$.

Proof.   This is again a proof by contradiction. We already know that $[E:F_G] \geq n$, so suppose $[E:F_G] \geq n + 1$. Then there are at least $n + 1$ elements $\alpha_1, \ldots, \alpha_{n+1}$ in E which are linearly independent over F_G. We can find $x_1, \ldots, x_{n+1} \in E$ not all zero so that

$$x_1\varphi_1(\alpha_1) + \cdots + x_{n+1}\varphi_1(\alpha_{n+1}) = 0,$$

$$x_1\varphi_2(\alpha_1) + \cdots + x_{n+1}\varphi_2(\alpha_{n+1}) = 0,$$

$$x_1\varphi_n(\alpha_1) + \cdots + x_{n+1}\varphi_n(\alpha_{n+1}) = 0,$$

because we need only satisfy n linear homogeneous equations with $n + 1$ unknowns. Suppose we let φ_1 be the identity automorphism, so that $\varphi_1(\alpha_i) = \alpha_i$. The solution for $x_1, \ldots, x_{n+1}$ will not generally be unique. Suppose we again choose one that contains the maximum possible number of 0's among the x's and that the x's are numbered in such a way that $x_1, \ldots, x_r \neq 0$, $x_{r+1} = \cdots = x_{n+1} = 0$. Clearly $r \neq 1$, because otherwise ———————————————————————————.
Also, since $x_r \neq 0$, we can divide through by x_r and get the new equations

$$a_1\varphi_1(\alpha_1) + \cdots + \varphi_1(\alpha_r) = 0, \left.\rule{0pt}{60pt}\right\} \qquad (7)$$
$$\vdots$$
$$a_1\varphi_n(\alpha_1) + \cdots + \varphi_n(\alpha_r) = 0,$$

2.4 FIXED FIELDS

where $a_i = x_i/x_r$. Not all $a_1, \ldots, a_{r-1}$ can be in F_G, for if they were, we would have $\varphi_j(a_i) = a_i$ for all i, j, and so

$$0 = a_1\varphi_j(\alpha_1) + \cdots + \varphi_j(\alpha_r)$$

$$= \varphi_j(a_1)\varphi_j(\alpha_1) + \cdots + \varphi_j(\alpha_r)$$

$$= \varphi_j(a_1\alpha_1 + \cdots + \alpha_r),$$

from which it follows that

$$a_1\alpha_1 + \cdots + \alpha_r = 0,$$

since φ_j is an isomorphism. But then $\alpha_1, \ldots, \alpha_r$ would not be linearly independent over F_G, so this cannot hold. Therefore, there must be at least one $i\,(1 \le i \le r - 1)$ and one $j\,(1 \le j \le n)$ for which $\phi_j(a_i) \ne a_i$. Choose one such j and keep it fixed. By hypothesis G is a group and so $\varphi_j\varphi_i \in G$ for every i and j. If we write φ_{k_i} for $\varphi_j\varphi_i$ we get from (7) the new equations

$$\varphi_j(a_1)\varphi_{k_1}(\alpha_1) + \cdots + \varphi_{k_1}(\alpha_r) = 0,$$

$$\vdots \tag{8}$$

$$\varphi_j(a_1)\varphi_{k_n}(\alpha_1) + \cdots + \varphi_{k_n}(\alpha_r) = 0,$$

Subtracting the ith equation of (8) from the k_ith of (7) gives

$$(a_1 - \varphi_j(a_1))\varphi_{k_1}(\alpha_1) + \cdots + (a_{r-1} - \varphi_j(a_{r-1}))\varphi_{k_1}(\alpha_{r-1}) + (\varphi_{k_1}(\alpha_r) - \varphi_{k_1}(\alpha_r)) = 0,$$

$$(a_1 - \varphi_j(a_1))\varphi_{k_2}(\alpha_1) + \cdots + (a_{r-1} - \varphi_j(a_{r-1}))\varphi_{k_2}(\alpha_{r-1}) = 0,$$

$$\vdots$$

$$(a_1 - \varphi_j(a_1))\varphi_{k_n}(\alpha_1) + \cdots + (a_{r-1} - \varphi_j(a_{r-1}))\varphi_{k_n}(\alpha_{r-1}) = 0.$$

We know that $a_i - \varphi_j(a_i) \ne 0$, for some i, so this last set of equations gives a new nontrivial solution for $x_1, \ldots, x_n$ in which we now have $x_r = x_{r+1} = \cdots = x_{n+1} = 0,$

$x_1 \neq 0$, so r could not have been minimal. We have reached a contradiction, so we must have $[E:F_G] = n$. $\|$

This theorem already proves part of the fundamental theorem of Galois theory. It has two important corollaries:

COROLLARY 1.   If G is a group of automorphisms of the field E and φ is any automorphism of E that leaves F_G fixed, then $\varphi \in G$.

Proof. If not, then F_G would remain fixed under the $(n + 1)$ automorphisms $\varphi_1, \ldots, \varphi_n, \varphi_{n+1}$, so $[E:F_G] \geq n + 1$, contradicting the theorem. $\|$

COROLLARY 2.   If G_1 and G_2 are two different finite groups of automorphisms of E, then $F_{G_1} \neq F_{G_2}$.

Proof.   Exercise.

EXAMPLE.   Let $E = Q(\omega, \sqrt[3]{2})$, let $\varphi_1, \ldots, \varphi_6$ be as in Section 2.3, p. 52, and let

$$G_1 = \{\varphi_1, \varphi_2\},$$

$$G_2 = \{\varphi_1, \varphi_4, \varphi_5\}.$$

Then G_1 is a group, since $\varphi_2^2(\sqrt[3]{2}) = \underline{\hspace{3cm}}$ and $\varphi_2^2(\omega) = \underline{\hspace{2cm}}$
$\underline{\hspace{3cm}}$, so $\varphi_2^2 = \varphi_1$. Also G_2 is a group, because

$$\varphi_4^2(\sqrt[3]{2}) = \underline{\hspace{3cm}} = \varphi_5(\sqrt[3]{2}),$$

$$\varphi_4^2(\omega) = \underline{\hspace{3cm}} = \varphi_5(\omega),$$

so $\varphi_4^2 = \varphi_5$. Moreover,

$$\varphi_4^3(\sqrt[3]{2}) = \varphi_4(\varphi_4^2(\sqrt[3]{2})) = \varphi_4(\underline{\hspace{2cm}}) = \underline{\hspace{2cm}},$$

$$\varphi_4^3(\omega) = \underline{\hspace{2cm}} = \underline{\hspace{2cm}} = \underline{\hspace{2cm}},$$

so

$$\varphi_4^3 = \underline{\hspace{3cm}}.$$

Therefore, $\varphi_4\varphi_5 = \varphi_4(\varphi_4^2) = \varphi_4^3 = \underline{\hspace{3cm}} = \varphi_5\varphi_4$. This shows that G_2 is a group.

First we want to find F_{G_1}, that is, all $\alpha \in Q(\sqrt[3]{2}, \omega)$ for which $\varphi_2(\alpha) = \alpha$. If $a, b, \ldots, f \in Q$ and

$$\alpha = a + b2^{1/3} + c2^{2/3} + d\omega + e\omega 2^{1/3} + f\omega 2^{2/3},$$

then

$$\varphi_2(\alpha) = \underline{\hspace{8cm}},$$

so $\varphi_2(\alpha) = \alpha$ implies that

$$b(2^{1/3} - \omega 2^{1/3}) + c(2^{2/3} - \omega^2 2^{2/3}) + d(\underline{\hspace{2cm}}) + e(\underline{\hspace{2cm}})$$

$$+ f(\underline{\hspace{2cm}}) = 0,$$

or equivalently (using the fact that $1 + \omega + \omega^2 = 0$),

$$d + (b - e)2^{1/3} + (2c)2^{2/3} + (\underline{\hspace{2cm}})\omega + (\underline{\hspace{2cm}})\omega 2^{1/3}$$

$$+ (\underline{\hspace{2cm}})\omega 2^{2/3} = 0.$$

From this we see that a is arbitrary, $b = e$, $c = 0$, $d = 0$, and f is also arbitrary. Therefore, α is of the form

$$\alpha = a + b2^{1/3} + b\omega 2^{1/3} + f\omega 2^{2/3} = a + (-b)\omega^2 2^{1/3} + f\omega 2^{2/3},$$

and so $F_{G_1} = Q(\omega^2 2^{1/3})$.

To find F_{G_2}, we now suppose that $\alpha = \varphi_4(\alpha) = \varphi_5(\alpha)$; that is,

$$a + b2^{1/3} + c2^{2/3} + d\omega + e\omega 2^{1/3} + f\omega 2^{2/3}$$

$$= a + b(\omega 2^{1/3}) + c(\omega^2 2^{2/3}) + d\omega + e(\underline{\hspace{2cm}}) + f(\underline{\hspace{2cm}})$$

$$= a + b(\omega^2 2^{1/3}) + c(\underline{\hspace{2cm}}) + d\omega + e(\underline{\hspace{2cm}}) + f(\underline{\hspace{2cm}}).$$

From these equations we see that $\underline{\hspace{1cm}}$ and $\underline{\hspace{1cm}}$ are arbitrary and $\underline{\hspace{1cm}} = \underline{\hspace{1cm}} = \underline{\hspace{1cm}} = \underline{\hspace{1cm}} = 0$. Therefore, $F_{G_2} = Q(\underline{\hspace{2cm}})$, and $F_{G_1} \neq F_{G_2}$.

CHAPTER III

FUNDAMENTAL THEOREM

3.1 SPLITTING FIELDS

A large part of Galois theory is concerned with the study of the automorphisms of the field generated by the roots of a given polynomial $p(x)$. Such a field is known as a splitting field of $p(x)$ (see also Table 3.1). More precisely, we have the following:

DEFINITION.   If $p(x)$ is a polynomial over F and $p(x)$ can be factored into linear factors in an extension E of F but cannot be factored into linear factors in any subfield of E, then E is called a *splitting field of $p(x)$* and one says "$p(x)$ splits in E."

For example, if $p(x) \in Q[x]$, and $p(x)$ has roots $\alpha_1, \ldots, \alpha_n$ in the field A of all algebraic numbers, then $E = Q(\alpha_1, \ldots, \alpha_n)$ is a splitting field of $p(x)$. More generally, we have

THEOREM.   If $p(x) \in F[x]$, then there is a field E that is a splitting field for $p(x)$.

Proof.   If $p(x) = p_1(x) \cdot \ldots \cdot p_k(x)$ and each $p_i(x)$ is irreducible over F, then $F_1 = F[x]/(p_1(x))$ will be a field in which $p_1(x)$ has a root α_1. So $p_1(x)$ factors into $(x - \alpha_1)p_{1,1}(x)$ over F_1, where $p_{1,1}(x)$ is a polynomial over F_1 and is either irreducible over F_1 or else is a product of irreducible factors. If $q_1(x)$ is an irreducible factor of $p_{1,1}(x)$, then $F_2 = F_1[x]/(q_1(x))$ is a field in which $q_1(x)$ has a root α_2 and so in F_2, $p_1(x)$ now has (at least) two linear factors. We continue this process until we reach a field F_i in which

69

3.1 SPLITTING FIELDS

TABLE 3.1 Examples of Splitting Fields

	$p(x) \in Q[x]$	E	Is E a splitting field of $p(x)$? If not, why not?
1	$x^2 - 5$	$Q(\sqrt{5})$	
2	$x^2 - 5$	$Q(\sqrt{-5})$	
3	$x^3 - 5$	$Q(\sqrt[3]{5})$	No, since $x^3 - 5 = (x - \sqrt[3]{5})(\underline{\hspace{4cm}})$ and the second factor is irreducible in E, as it has no real roots and E is real.
4	$x^3 - 5$	$Q(\sqrt[3]{5}, \omega\sqrt[3]{5})$	
5	$x^3 - 5$	$Q(\sqrt[3]{5}, \omega\sqrt[3]{5}, \omega^2\sqrt[3]{5})$	
6	$x^3 - 5$	$Q(\sqrt[3]{5}, \sqrt{-3})$	
7	$x^3 - 5$	R	
8	$x^3 - 5$	$Q(\sqrt[3]{5}, \sqrt{-3}, i)$	
9	$x^3 - 5$	C	
10	$x^2 + 1$	C	

$p_1(x)$ factors into linear factors. Then let $q_2(x)$ be an irreducible factor of $p(x)$ in F_i and continue the process until we reach a field E in which $p(x)$ splits. ‖

Given a polynomial $p(x) \in F[x]$, is the splitting field of $p(x)$ uniquely determined or can there be several different splitting fields, as the wording of the definition perhaps suggests? If you have examined the proof of the last theorem, you will see that this is certainly the case and a given $p(x)$ may have several different splitting fields, for the factors could have been presented in a different order and we could also have made use of the isomorphism $F_i(\gamma) \cong F_i[x]/(r(x))$, where $r(x)$ is irreducible over F_i and $r(\gamma) = 0$, at several stages. For example, $Q(\sqrt{2}, \sqrt{3})$ and $Q(\sqrt{2})[x]/(x^2 - 3)$ are both splitting fields of $p(x) = (x^2 - 2)(x^2 - 3)$. Examples include the following:

$p(x)$	F	E	Splitting of $p(x)$
$x^2 - 2$	Q	$Q(\sqrt{2})$	$(x - \sqrt{2})(x + \sqrt{2})$
		$Q[z]/(z^2 - 2)$	$(x - z)(x + z)$
$x^3 - 2$	Q	$Q(\omega, \sqrt[3]{2})$	$(x - \sqrt[3]{2})(x - \omega\sqrt[3]{2})(x - \omega^2\sqrt[3]{2})$
		$Q(\omega)[z]/(z^3 - 2)$	$(x - z)(x - \omega z)(x - \omega^2 z)$
		$Q(\sqrt[3]{2})[y]/(y^2 + y + 1)$	$(x - \sqrt[3]{2})(x - y\sqrt[3]{2})(x - y^2\sqrt[3]{2})$

Fortunately we do, however, have the

THEOREM. All splitting fields of a given polynomial are isomorphic [A, p. 31].

This theorem will follow immediately from Kronecker's theorem on extensions of isomorphisms, which is proved later in this section.

3.1 SPLITTING FIELDS

EXAMPLE. The fields $E_1 = Q(\sqrt{2})$ and $E_2 = Q[z]/(z^2 - 2)$ are clearly isomorphic by the isomorphism $\varphi: \sqrt{2} \to z$. Is there another isomorphism of E_1 onto E_2?

For each of the following fields, is there a polynomial $p(x)$ for which it is the splitting field of $p(x)$ over Q?

Q	R	$Q(\sqrt[3]{5})$	$Q(\sqrt{5})$	$Q(\sqrt{5}, \sqrt[3]{5})$	$Q(\sqrt{2}, \sqrt{5})$	$Q(\sqrt[3]{2}, \sqrt{-3})$

In order to prove that all splitting fields of a given polynomial are isomorphic, we shall need a theorem on the existence of extensions of isomorphisms from subfields to all of E. The following example shows why such a theorem is necessary.

Suppose that we have four fields F, F', K, and K', where $F \subset K$, $F' \subset K'$, and $K \cong K'$ (see Figure 3.1). There is a great temptation to believe that any isomorphism φ of F onto F' can be extended to an isomorphism of K onto K'. This is not true, however. For example, let

$$F = Q(\sqrt{2}), \qquad K = Q(2^{1/4}),$$
$$F' = Q(-\sqrt{2}), \qquad K' = Q(2^{1/4}).$$

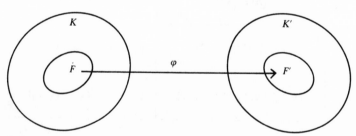

FIGURE 3.1

Here F is actually the same as F' and K the same as K', so of course $K \cong K'$.

Now let φ be the isomorphism

$$\varphi: a + b\sqrt{2} \to a - b\sqrt{2}.$$

Then φ cannot be extended to all of K: For if ψ is any isomorphism of K onto K'

and α is any element of K, then any statement true of α in K must also hold of $\alpha' = \psi(\alpha)$ in K'. So if α has a square root in K, then $\alpha' = \psi(\alpha)$ must have a square root in K'. But if ψ were an extension of φ, then ψ and φ would have the same effect on elements of F, and we would have for $\alpha = \sqrt{2}$ that

$$\alpha' = \psi(\sqrt{2}) = \varphi(\sqrt{2}) = -\sqrt{2}.$$

The isomorphism ψ would therefore transform the element $\alpha = \sqrt{2}$, which does have a square root in K, into the element $\alpha' = -\sqrt{2}$, which does not have a square root in K', since $K = K' = Q(2^{1/4})$. So φ cannot be extended to an automorphism of K.

Nonetheless, K does have a nontrivial automorphism, but it is the identity when restricted to F. For example, we may take $\psi(2^{1/4}) = -2^{1/4}$.

Can you construct another example of a mapping that cannot be extended? Can you find a field E such that in the example above, $F \subset K \subset E$, and every isomorphism φ of F onto F' can be extended to E?

Kronecker considered these questions and arrived at the following result:

THEOREM. If

(1) φ is an isomorphism of F onto F', and

(2) under this isomorphism φ, the polynomial $f(x) \in F[x]$ corresponds to the polynomial $f'(x) \in F'[x]$, and

(3) E is a splitting field of $f(x)$, and E' of $f'(x)$,

then φ can always be extended to an isomorphism of E and E' (see Figure 3.2).

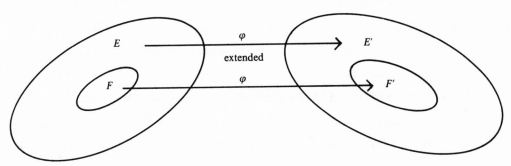

FIGURE 3.2

3.1 SPLITTING FIELDS

EXAMPLE

F	F'	$f(x)$	$f'(x)$	$E = E'$ (in this case)
$Q(\sqrt[3]{5})$	$Q(\omega\sqrt[3]{5})$	$x^2 + \sqrt[3]{5}x + 5^{2/3}$	$x^2 + \omega\sqrt[3]{5}x + \omega^2 5^{2/3}$	$Q(\sqrt[3]{5}, \omega)$

Let $\varphi : Q(\sqrt[3]{5}) \to Q(\omega\sqrt[3]{5})$ be determined by $\varphi : \sqrt[3]{5} \to \omega\sqrt[3]{5}$. Then φ can be extended to E by the definition

$$\varphi : \begin{cases} \sqrt[3]{5} \to \omega\sqrt[3]{5} \\ \omega \to \begin{cases} \omega \\ \text{or} \\ \omega^2, \end{cases} \end{cases}$$

whichever you prefer. Therefore, the extension of φ to E is not necessarily unique; the theorem only says that there exists at least one such extension.

Proof of the Theorem. The proof is by induction on k, where k is the number of those roots of $f(x)$ which are not in F.

If $k = 0$, then all the roots of $f(x)$ are in F, so $f(x)$ must split in F and we have $E = F$. Therefore, $f'(x)$ must split in F' and $E' = F'$. The original isomorphism φ is therefore also an isomorphism of E onto E'.

Next we suppose that the theorem holds whenever the number of roots not in the ground field is at most $k - 1$ and that $f(x)$ has k roots outside F. If we factor $f(x)$ into irreducible polynomials over F we get $f(x) = f_1(x) \cdot \ldots \cdot f_r(x)$ and a corresponding factorization $f'(x) = f_1'(x) \cdot \ldots \cdot f_r'(x)$ of $f'(x)$ over F' under the isomorphism φ. At least one of the factors $f_1, \ldots, f_r$ must be of degree higher than 1 because _____

_____. Suppose it is $f_1(x)$ and that α is a root of f_1. Then $f_1'(x) = \varphi(f_1(x))$ is also of degree higher than 1 because _____

_____. Let α' be a root of $f_1'(x)$. Then writing a_i' for $\varphi(a_i)$ when $a_i \in F$, we have $F(\alpha) \cong F[x]/(f_1(x))$ by the isomorphism

$$a_0 + a_1\alpha + \cdots + a_s\alpha^s \longleftrightarrow a_0 + a_1x + \cdots + a_sx^s + (f_1(x)).$$

Also, $F[x]/(f_1(x)) \cong F'[x]/(f'_1(x))$ by the isomorphism

$$a_0 + \cdots + a_s x^s + (f_1(x)) \longleftrightarrow a'_0 + \cdots + a'_s x^s + (f'_1(x)).$$

and $F'[x]/(f'_1(x)) \cong F'(\alpha')$ by the isomorphism

$$a'_0 + \cdots + a'_s x^s + (f'_1(x)) \longleftrightarrow a'_0 + \cdots + a'_s(\alpha')^s.$$

Combining these isomorphisms gives $F(\alpha) \cong F'(\alpha')$ by the isomorphism

$$a_0 + \cdots + a_s \alpha^s \leftrightarrow a'_0 + \cdots + a'_s(\alpha')^s.$$

This last isomorphism which we shall call ψ is clearly an extension of φ because

———.

We now take $F(\alpha)$ as out new ground field. Then $f(x)$ has only $k - 1$ roots outside $F(\alpha)$, and $f'(x)$ has only $k - 1$ roots outside $F'(\alpha')$. Therefore, ψ has an extension χ to an isomorphism of E to E' by our induction hypothesis. Since χ is an extension of ψ and ψ is an extension of φ, χ is also an extension of φ, which proves the theorem. ‖

COROLLARY. All splitting fields of a polynomial $p(x) \in F[x]$ are isomorphic.

Proof. Let $F' = F$ and let K and K' both be splitting fields of $p(x)$. Then $K \cong K'$, by the theorem. ‖

COROLLARY. All finite fields containing p^n elements are isomorphic.

Proof. If F is any field containing p^n elements, then the multiplicative group of nonzero elements of F contains $p^n - 1$ elements. The order of an element is always a divisor of the order of the group, so each nonzero element satisfies the equation $x^{p^n - 1} = 1$. Therefore, each of the p^n different elements of F satisfies $x^{p^n} - x = 0$ and F is the splitting field of this polynomial. By the theorem, all such fields are isomorphic. ‖

3.2 NORMAL EXTENSIONS AND GROUPS OF AUTOMORPHISMS

We have already seen that the set of all automorphisms of a field E is a group with the composition of mappings as group operation. If now $F \subset E$ and φ_1, φ_2 are two automorphisms of E which leave F fixed, then $\varphi_1 \varphi_2$ is also an automorphism of E which leaves F fixed, and so is φ_1^{-1}. The identity mapping of E onto itself of course always leaves F fixed. So we see that the set of all those automorphisms of E which leave F fixed is also a group.

DEFINITION. If $F < E$, then the *group of E over F* is the group consisting of all those automorphisms of E which leave every element of F fixed. This group is usually denoted by $G(E/F)$.

Notice that therefore $\varphi \in G(E/F)$ if and only if $\varphi \restriction F = I$. [By $\varphi \restriction F$ we mean the mapping φ' which is obtained from φ by limiting its domain to F, setting $\varphi'(x) = \varphi(x)$ for every $x \in F$. I is the identity mapping.]

EXAMPLE. Let $E = Q(\sqrt[3]{2}, \sqrt[3]{5}, \omega)$. This field has 18 automorphisms, as given in Table 3.2.

From this table we see that

$G(E/Q(\sqrt[3]{2}, \sqrt[3]{5})) = \{\varphi_1, \varphi_2\}$, and $[E:Q(\sqrt[3]{2}, \sqrt[3]{5})] = \underline{\hspace{1cm}}$,

$G(E/Q(\sqrt[3]{5})) = \{\underline{\hspace{7cm}}\}$

and $[E:Q(\sqrt[3]{5})] = \underline{\hspace{1cm}}$,

$G(E/Q(\omega)) = \{\underline{\hspace{7cm}}\}$

and $[E:Q(\omega)]_3 = \underline{\hspace{1cm}}$,

$G(E/Q(\omega\sqrt[3]{2})) = \{\underline{\hspace{6cm}}\}$

and $[E:Q(\omega\sqrt[3]{2})] = \underline{\hspace{1cm}}$,

$G(E/Q(\omega^2\sqrt[3]{5}, \omega)) = \{\underline{\hspace{6cm}}\}$

and $[E:Q(\omega^2\sqrt[3]{5}, \omega)] = \underline{\hspace{1cm}}$.

TABLE 3.2

	$\sqrt[3]{2}\rightarrow$	$\sqrt[3]{5}\rightarrow$	$\omega\rightarrow$
φ_1	$\sqrt[3]{2}$	$\sqrt[3]{5}$	ω
φ_2	$\sqrt[3]{2}$	$\sqrt[3]{5}$	ω^2
φ_3	$\sqrt[3]{2}$	$\omega\sqrt[3]{5}$	ω
φ_{18}			

3.2 NORMAL EXTENSIONS AND GROUPS OF AUTOMORPHISMS

Now suppose that E is a field and $F < E$. We can then form $G(E/F)$ and call it Γ. Next we look at Γ and try to find the field F_Γ of elements left fixed by Γ. How is F_Γ related to F? Clearly, $F \le F_\Gamma$, because every $\varphi \in \Gamma$ leaves F fixed. It is, however, conceivable that φ might leave fixed some field larger than F. Does this case ever occur? That is, can we ever have $F \subsetneq F_\Gamma$? Or do we always have $F = F_\Gamma$? Again we look first at specific examples.

EXAMPLES.   (1) Let $E = Q(\sqrt[3]{2}, \sqrt[4]{5}, i)$, $F = Q(i)$. The 8 automorphisms of E are

	$i \rightarrow$	$\sqrt[3]{2} \rightarrow$	$\sqrt[4]{5} \rightarrow$
φ_1	i	$\sqrt[3]{2}$	$\sqrt[4]{5}$
φ_2	$-i$	$\sqrt[3]{2}$	$\sqrt[4]{5}$
φ_3	i	$\sqrt[3]{2}$	$i\sqrt[4]{5}$
φ_4			
φ_5			
φ_6			
φ_7			
φ_8	$-i$	$\sqrt[3]{2}$	$-i\sqrt[4]{5}$

This table shows that $\Gamma = G(E/F) = \{\varphi_1, \underline{\hspace{4cm}}\}$. However, each of these automorphisms leaves fixed every element of $Q(i, \sqrt[3]{2})$, so $F_\Gamma = Q(i, \sqrt[3]{2})$ and $F \neq F_\Gamma$.

(2) Let $E = Q(\sqrt[3]{2}, \sqrt[4]{5}, i)$, $F = Q(\sqrt[3]{2})$. Then $\Gamma = G(E/F) = \{\underline{\hspace{3cm}}$

$\underline{\hspace{2cm}}\}$, and $F_\Gamma = \underline{\hspace{4cm}}$, so in this case $F\begin{Bmatrix} = \\ \neq \end{Bmatrix} F_\Gamma$ (cross out one).

(3) Let $E = Q(\sqrt[3]{5}, \omega)$, $F = Q(\omega)$. The automorphisms of $= G(E/F)$ are

	$\sqrt[3]{5} \rightarrow$	$\omega \rightarrow$
ϕ_1	$\sqrt[3]{5}$	ω
ϕ_2		ω
ϕ_3		ω

and so $F_\Gamma \begin{Bmatrix} = \\ \neq \end{Bmatrix} F$.

(Incidentally, you may have the impression from these examples that it does not affect the group of a field whether its definition used $\sqrt[3]{2}$, $\sqrt[3]{5}$, or some roots of still other primes. This is often true, but not always, so as usual, care is necessary: For example if ε is a primitive fourth root of unity and $F_1 = Q(\varepsilon, \sqrt{2})$, $F_2 = Q(\varepsilon, \sqrt{3})$, then $[F_1 : Q] = \underline{\hspace{1.5cm}}$, while $[F_2 : Q] = \underline{\hspace{1.5cm}}$, since $\sqrt{2} \in Q(\varepsilon)$, while $\sqrt{3} \notin Q(\varepsilon)$.)

3.2 NORMAL EXTENSIONS AND GROUPS OF AUTOMORPHISMS

Those extensions E of F for which we do have $F = F_\Gamma$, where $\Gamma = G(E/F)$, are of special importance in the one-to-one correspondence between groups and fields which will be set up by the fundamental theorem of Galois theory. They are called *normal extensions*:

DEFINITION. A finite extension E of F is called a *normal* extension of F if $G(E/F) = \Gamma$ implies that $F_\Gamma = F$. We write $F \lhd E$ or $E \rhd F$.

The preceding examples show that $Q(\sqrt[3]{2}, \sqrt[4]{5}, i)$ is not a normal extension of $Q(i)$, but that it is a normal extension of $Q(\sqrt[3]{2})$. They show also that $Q(\sqrt[3]{5}, \omega)\begin{cases} \text{is} \\ \text{is not} \end{cases}$ a normal extension of $Q(\omega)$.

Before stating a theorem that gives a necessary and sufficient condition that E be a normal extension of F, we shall give a few more examples (see Table 3.3).

First we have the following theorem on normal extensions:

THEOREM. Suppose the irreducible polynomial $p(x)$ has coefficients in F and that $E \rhd F$. If E contains one root α of $p(x)$, then E contains all the roots of $p(x)$ and these roots are all different from each other.

Proof. Let $\alpha_1, \ldots, \alpha_k$ be the different conjugates of α in E under $G(E/F)$; that is,

$$\{\alpha_1, \ldots, \alpha_k\} = \{\gamma | \gamma = \sigma(\alpha) \text{ for some } \sigma \in G(E/F), \text{ and } \alpha_i \neq \alpha_j \text{ for } i \neq j\}.$$

We shall suppose that $\alpha = \alpha_1$.

Since α is a root of $p(x)$, we know that $p(x)$ factors into

$$p(x) = (x - \alpha)q(x)$$

over E. If $\sigma \in G(E/F)$, then for some integer i we have

$$p(x) = \sigma(p(x)) = (x - \sigma(\alpha))\sigma(q(x))$$

$$= (x - \alpha_i)\sigma(q(x)),$$

TABLE 3.3

	E	F	$G(E/F) = \Gamma$	Fixed field F_Γ	$[E:F]$	$[E:F_\Gamma]$	Remarks
1	$Q(i)$	Q	$\sigma : i \to \begin{cases} +i \\ -i \end{cases}$ (2 elements)	Q	2	2	Normal
2	$Q(\sqrt[3]{2})$	Q					
3	$Q(\sqrt[3]{2}, \omega)$	Q	$\sigma : \sqrt[3]{2} \to \begin{cases} \sqrt[3]{2} \\ \sqrt[3]{2}\omega \\ \sqrt[3]{2}\omega^2 \end{cases} \quad \omega \to \begin{cases} \omega \\ \omega^2 \end{cases}$				
4	$Q(\sqrt[6]{2}, \varepsilon_6)$	Q	$\sigma : \sqrt[6]{2} \to \begin{cases} \sqrt[6]{2} \end{cases} \quad \varepsilon \to \begin{cases} \varepsilon \\ \varepsilon^5 \end{cases}$ (12 elements)				
5	$Q(\sqrt[8]{3}, i)$ (real eighth root)	Q	(8 elements)				Not normal
6	$Q(\sqrt[8]{3}, \varepsilon_8)$	Q	(32 elements)				
7	$Q(\sqrt[8]{3}, \varepsilon_8)$	$Q(i)$					
8	$Q(\sqrt[5]{3}, i)$	$Q(i)$					

Note: In all cases ε_N denotes a primitive nth root of unity, so $\varepsilon^N = 1$, but $\varepsilon^k \neq 1$ for $1 \leq k \leq N - 1$.

3.2   NORMAL EXTENSIONS AND GROUPS OF AUTOMORPHISMS

where as usual we write $\sigma(f(x))$ for $\sigma(c_0) + \sigma(c_1)x + \cdots + \sigma(c_r)x^r$ whenever $f(x) = c_0 + \cdots + c_r x^r$, so that $p(x) = \sigma(p(x))$ because _____

_____.

Therefore each $(x - \alpha_i)$ is a factor of $p(x)$ in E and so $p(x)$ must factor into

$$p(x) = (x - \alpha_1)(x - \alpha_2)\cdots(x - \alpha_k) \cdot r(x)$$

over E.

The automorphisms $\sigma \in G(E/F)$, however, only permute the factors of $s(x) = (x - \alpha_1)\cdots(x - \alpha_k)$, so $\sigma\,(p(x)) = \sigma(s(x)) \cdot \sigma(r(x))$ and we see that $\sigma(r(x)) = r(x)$ for every $\sigma \in G(E/F)$.

Both $s(x)$ and $r(x)$ are therefore polynomials in $F[x]$. As a consequence, $r(x)$ must be a constant c, for _____

_____, and so $p(x) = cs(x) = c(x - \alpha_1)\cdots(x - \alpha_k)$ and its roots are all different from each other. $\|$

Next we obtain a very important characterization of normal extensions:

THEOREM.   The field E is a normal extension of F if and only if E is the splitting field of a polynomial $p(x)$ which has coefficients in F and no repeated roots.

Proof.   (1) First we suppose that $E \rhd F$. Then $[E:F]$ is finite, because _____

_____, so suppose that $\gamma_1, \gamma_2, \ldots, \gamma_h$ are a basis for E over F. If γ_1 is a root of some polynomial $p_1(x), \gamma_2$ a root of $p_2(x), \ldots,$ and γ_h a root of $p_h(x)$, where all $p_i(x) \in F[x]$ and are irreducible over F, then E is the splitting field of $p(x) = \text{lcm}(p_1(x), p_2(x), \ldots, p_h(x))$, because by the preceding theorem _____.

Moreover, this polynomial has no repeated roots, because _____

_____.

(2) Now we suppose that E is the splitting field of $p(x) \in F[x]$, where $p(x)$ has no repeated roots. We must prove that $E \rhd F$.

First, we notice immediately that $[E:F]$ is finite, because _____

_____.

We carry out the proof by induction on the degree of the splitting field E over the ground field K. The induction hypothesis will be that whenever E is the splitting field of $p(x)$, where $p(x) \in K[x]$ and has no repeated roots, and $[E:K] \leq n - 1$, then $E \vartriangleright K$. If $n - 1 = 1$, this is surely true. Suppose now that E is the splitting field of $p(x) = p_1(x) \cdot \ldots \cdot p_r(x) \in F[x]$, where $p_1(x)$ is irreducible of degree $k > 1$ with roots $\alpha_1, \ldots, \alpha_k$ and that $[E:F] = n$. Then $[F(\alpha_1):F] = k > 1$, so if we let $K = F(\alpha_1)$, then $[E:K] = [E:F]/[K:F] = n/k \leq n - 1$, and E is also the splitting field of $p(x)$ over K. By the induction hypothesis, therefore, $E \vartriangleright K$. So we know that whenever $\gamma \in E$ and $\gamma = \varphi(\gamma)$ for every $\varphi \in G(E/K)$, then $\gamma \in K$ and must thus be of the form

$$\gamma = a_0 + a_1\alpha_1 + \cdots + a_{k-1}\alpha_1^{k-1} \tag{1}$$

with $a_i \in F$. We must still show that $E \vartriangleright F$; that is, we must show that _____

_____.

So suppose $\gamma \in E$ and $\gamma = \sigma(\gamma)$ for every $\sigma \in G(E/F)$. Every automorphism that leaves K fixed will also leave F fixed, so γ is left fixed by every $\varphi \in G(E/K)$ and is of the form of equation (1). The polynomial $p_1(x)$ has no repeated roots and $F(\alpha_1) \cong F(\alpha_i)$ for every $i = 1, \ldots, k$, because _____

_____. Therefore, for every $i = 1, \ldots, k$ there is an automorphism σ_i of E leaving F fixed such that $\sigma_i(\alpha_1) = \alpha_i$. Then

$$\gamma = \sigma_i(\gamma) = \sigma_i(a_0 + a_1\alpha_1 + \cdots + a_{k-1}\alpha_1^{k-1})$$

$$= a_0 + a_1\alpha_i + \cdots + a_{k-1}\alpha_i^{k-1}. \tag{2}$$

Now let $r(x) = (\gamma - a_0) + a_1x + \cdots + a_{k-1}x^{k-1}$. This is a polynomial of degree $k - 1$ over the field K with k different roots $\alpha_1, \ldots, \alpha_k$, by equation (2). However, the only polynomial of degree $k - 1$ that has more than $k - 1$ roots is the zero

3.2 NORMAL EXTENSIONS AND GROUPS OF AUTOMORPHISMS

polynomial. Therefore, $a_{k-1} = a_{k-2} = \cdots = a_1 = (\gamma - a_0) = 0$, and since $a_0 \in F$ we must have $\gamma \in F$, which is just what we wanted. ‖

There is another characterization of normal extensions:

THEOREM. $E \rhd F \Leftrightarrow |G(E/F)| = [E:F]$.

Proof. If F_G is the fixed field of $G(E/F)$, then $|G(E/F)| = [E:F_G]$, because _____

_____ . Therefore,

$$E \rhd F \Leftrightarrow F = F_G, \qquad \text{by the definition of normality}$$

$$\Leftrightarrow [E:F] = [E:F_G] = |G(E/F)|. \qquad ‖$$

This theorem was the motivation for including the columns with degrees in preceding exercises (Tables 2.10 and 3.3).

Notice that the splitting field E of $f(x) = p_1^{n_1}(x) \cdot \ldots \cdot p_k^{n_k}(x)$, is also the splitting field of $g(x) = p_1(x) \cdot \ldots \cdot p_k(x)$, so repetitions of any factor $p_i(x)$ of $f(x)$ have no effect on E and its automorphisms. It is, however, possible for $f(x)$ to be irreducible over F, yet to have repeated roots in E.

EXAMPLE. Evidently $f(x)$ has a repeated root α if $f(x) = (x - \alpha)^2 g(x)$, when factored over E. In this case

$$f'(x) = \frac{d}{dx} f(x) = \text{\underline{\hspace{6cm}}},$$

and so $f(x)$ and $f'(x)$ have $(x - \alpha)$ as a gcd, unless perhaps $f'(x)$ vanishes identically.

Suppose now that $f(x)$ is irreducible over F but not a constant. Then f and f' cannot have a common factor, because f' is of lower degree, so we cannot have a repeated root unless $f'(x) = 0$, identically. This cannot happen in a field of char 0, because _____,

but if char $F = p \neq 0$, this is easily possible. A simple example is the following:

Let $F_2 = GF(2)$, the field consisting of only the two elements 0, 1 under $+$, $\cdot$ (mod 2), and let $F = F_2(z)$, the field of rational functions in z with coefficients in F_2. Now examine the polynomial $f(x) = x^2 + z$. This polynomial is irreducible over F, for, if it were reducible, it could only have linear factors: $f(x) = x^2 + z = (ax + b)(cx + d) = acx^2 + (ad + bc)x + bd$, with $a, b, c, d \in F$. We may take $a = 1$, so that automatically $c = 1$. Then $ad + bc = d + b = 0$ and $bd = z$. This gives $d = -b = b$ (because we are working mod 2) and $bd = b^2 = z$, so $b = \sqrt{z}$. But $\sqrt{z} \notin F$, so $f(x)$ is irreducible over F. However, in the extension field $E = F(\sqrt{z})$ we have

$$f(x) = x^2 + z = x^2 + (2\sqrt{z})x + z = (x + \sqrt{z})^2,$$

which has the repeated root $\sqrt{z}$, in spite of the fact that $x^2 + z$ is irreducible over F. Note incidentally that here

$$\frac{d}{dx} f(x) = \underline{\hspace{3cm}} = 0 \text{ in } F.$$

In this particular example we find that $G(E/F)$ is _____, so the field F_0 fixed by this group is $F_0 = $ _____, and $[E:F] = $ _____, while $|G(E/F)| = $ _____. This shows that the hypothesis "no repeated roots in E" is really necessary.

Polynomials without repeated roots are therefore of special importance:

DEFINITION. A polynomial $f(x) \in F[x]$ is called *separable* if none of its irreducible factors has a repeated root.

DEFINITION. If $F \subseteq E$ and $\alpha \in E$, then α is called *separable over* F if it is the root of a separable polynomial over F.

3.2 NORMAL EXTENSIONS AND GROUPS OF AUTOMORPHISMS

DEFINITION. If $F \subseteq E$, then E is called a *separable extension of F* if every element $\alpha \in E$ is separable over F. (E need not be a finite extension.)

There is a discussion of separable and inseparable extensions in [vdW, pp. 119–125]. We shall now define perfect fields and prove a theorem that characterizes them.

DEFINITION. A field F is called *perfect* if every irreducible polynomial $p(x) \in F[x]$ is separable. Otherwise it is called *imperfect*.

THEOREM. (1) If char $F = 0$, then F is perfect.

(2) If char $F = p$, then

F is perfect $\Leftrightarrow$ every element of F has a pth root which is also in F.

Proof. (1) If $p(x)$ is irreducible, then deg $p(x) = n \geq 1$. Therefore, the derivative $p'(x) \not\equiv 0$. However, if $p(x)$ has a repeated root α, then $(x - \alpha)^2 | p(x)$, so

$$p(x) = (x - \alpha)^2 \cdot r(x),$$

$$0 \not\equiv p'(x) = 2(x - \alpha)r(x) + (x - \alpha)^2 r'(x)$$

$$= (x - \alpha)s(x).$$

So $p(x)$ and $p'(x)$ have a factor $d(x)$ in common and we know that $d(x)$ can be expressed as a linear combination of $p(x)$ and $p'(x)$ using the Euclidean algorithm. Since deg $p'(x) = n - 1$ and n must be at least 2 in order to have a repeated root, so the degree of $d(x)$ must be at least 1 and at most $n - 1$, and $d(x)|p(x)$. But $p(x)$ was assumed to be irreducible, so we reach a contradiction showing that $p(x)$ cannot have repeated roots.

(2) We now have char $F = p$. We again suppose that

$$p(x) = (x - \alpha)^2 r(x) = c_0 + c_1 x + \cdots + c_n x^n,$$

and, as before,

$$p'(x) = (x - \alpha)s(x) = c_1 + \cdots + nc_n x^{n-1}.$$

If $p'(x) \not\equiv 0$, we reach a contradiction just as under case (1). However, if char $F = p$, we can have $p'(x) \equiv 0$; in fact, this happens whenever $p(x)$ contains only powers of x^p, for the derivative of x^p is px^{p-1}, which vanishes mod p. Should $p(x)$ therefore contain only powers of x^p, then $p'(x) \equiv 0$ and $p(x)$ might perhaps have repeated roots. If, however, every element of F has a pth root in F, then

$$p(x) = c_0 + c_p x^p + \cdots + c_{kp} x^{kp}$$

$$= (\sqrt[p]{c_0} + \sqrt[p]{c_1} x + \cdots + \sqrt[p]{c_{kp}} x^k)^p.$$

So $p(x)$ would be reducible, contrary to the hypothesis. Therefore, an irreducible polynomial cannot have repeated roots and F must be perfect.

Now suppose that some element α does not have a pth root in F, say $\beta = \sqrt[p]{\alpha} \notin F$. We wish to show that in this case F is imperfect. To do this, we shall show that the polynomial $p(x) = x^p - \alpha$ is irreducible over F. It has repeated roots, for

$$p(x) = x^p - \alpha = x^p - (\sqrt[p]{\alpha})^p = (x - \beta)^p \qquad \text{mod } p,$$

because $\binom{p}{k}$ is divisible by p for $k = 1, \ldots, p - 1$. Suppose $p(x)$ were reducible: Any irreducible factor $r(x)$ must be of the form $(x - \beta)^k$ for some $k \geq 1$. We cannot have $k = 1$, for this would imply that $\beta \in F$. Therefore, $r(x)$ must be divisible by $(x - \beta)^2$ and so have a repeated root. This makes $p(x)$ inseparable and F imperfect. If F is therefore to be perfect, then every element of F must have a pth root in F. $\quad \parallel$

Clearly, therefore, all algebraically closed fields are perfect. We shall show in Section 4.11 that all finite fields are perfect also.

To conclude this section we now define Galois groups:

DEFINITION.   If E is a normal extension of K, then $G(E/K)$ is called the *Galois group* of E over K.

DEFINITION.   By the *Galois group* of a polynomial $p(x) \in F[x]$ we mean $G(E/F)$, where E is the splitting field of $p(x)$.

3.3   CONJUGATE FIELDS AND ELEMENTS

EXAMPLE.   If $F = Q$ and $p(x) = x^3 - 2$, then the Galois group of $p(x)$ consists of the automorphism $\varphi_1, \ldots, \varphi_6$ (see Section 2.3) of $E = Q(\sqrt[3]{2}, \omega)$.

There will be more examples later, for this concept is Galois's great contribution and of course basic to Galois theory.

3.3   CONJUGATE FIELDS AND ELEMENTS

Let E be a normal extension of F and let K_1 be some intermediate field, $F < K_1 < E$. If the automorphism σ of E leaves every element of F fixed (so $\sigma|F = I$, the identity mapping on F), then in general σ is, of course, not the identity on K_1. It will, however, map K_1 either onto itself or onto some other subfield K_2 of E. Such a field is called conjugate to K_1 in E relative to F as the fixed field.

DEFINITION.   The field K_2 is conjugate to K_1 in the finite extension E of F if there is an automorphism σ of E such that $\sigma{\restriction}F = I$ and $\sigma K_1 = K_2$.

DEFINITION.   The element α_2 of E is conjugate to the element α_1 of E if there is an automorphism σ of E such that $\sigma{\restriction}F = I$ and $\sigma(\alpha_1) = \alpha_2$.

(If F is not explicitly specified, it is understood to be the field of those elements which are fixed by every automorphism of E, that is, $F = \{\alpha | \forall \sigma : \sigma\alpha = \alpha$, where σ is an automorphism of $E\}$.) Clearly, conjugacy is an equivalence relation.

EXAMPLES.   Here E is the splitting field of the polynomial $p(x) = (x^3 - 2)(x^3 - 5)$, so $E = Q(\omega, \sqrt[3]{2}, \sqrt[3]{5})$. There are 18 automorphisms σ:

$$\sqrt[3]{2} \to \begin{cases} \sqrt[3]{2} \\ \omega\sqrt[3]{2} \quad or \\ \omega^2\sqrt[3]{2} \end{cases} or \qquad \sqrt[3]{5} \to \begin{cases} \sqrt[3]{5} \\ \omega\sqrt[3]{5} \quad or \\ \omega^2\sqrt[3]{5} \end{cases} or \qquad \omega \to \begin{cases} \omega \\ or \\ \omega^2. \end{cases}$$

We consider several different fixed fields F and different possible fields K_1:

	Fixed field F	K_1	Conjugate fields to K_1 in E
1	Q	$Q(\sqrt[3]{2})$	$Q(\omega\sqrt[3]{2}), Q(\omega^2\sqrt[3]{2}), Q(\sqrt[3]{2}).$
2	Q	$Q(\omega, \sqrt[3]{2})$	$Q(\omega, \sqrt[3]{2}) = Q(\omega, \omega\sqrt[3]{2}) = Q(\omega, \omega^2\sqrt[3]{2})$
3	$Q(\omega)$	$Q(\omega, \sqrt[3]{2})$	$Q(\omega, \sqrt[3]{2})$
4	$Q(\omega)$	$Q(\omega, \sqrt[3]{2}, \sqrt[3]{5})$	
5	Q	$Q(\sqrt[3]{2}, \sqrt[3]{5})$	
6	$Q(\sqrt[3]{2})$	$Q(\sqrt[3]{2}, \sqrt[3]{5})$	

Notice that in example 2 the field K_1 is mapped onto itself by every automorphism σ of E that is, $\sigma \restriction K_1$ is always an automorphism of K_1, but that in general, $\sigma \restriction K_1 \neq I$.

The following theorem describes the correspondence between conjugate subfields of E and their Galois groups:

THEOREM. If E is a finite normal extension of F and $F < K_1$, $K_2 < E$, then K_1 is conjugate to $K_2 \Leftrightarrow G(E/K_1)$ and $G(E/K_2)$ are conjugate subgroups of $G(E/F)$.

Proof. (1) First suppose that K_1 is conjugate to K_2. If σ is an automorphism of E which carries K_1 onto K_2, then for every $\alpha_2 \in K_2$ there is some $\alpha_1 \in K_1$ such that $\alpha_2 = \sigma\alpha_1$. Now take any $\varphi \in G(E/K_1)$. Then $\varphi\alpha_1 = \alpha_1$, and so

$$\alpha_2 = \sigma\alpha_1 = \sigma(\varphi\alpha_1) = \sigma(\varphi\sigma^{-1}\alpha_2) = (\sigma\varphi\sigma^{-1})\alpha_2.$$

3.3 CONJUGATE FIELDS AND ELEMENTS

Therefore, $\sigma\varphi\sigma^{-1} \in G(E/K_2)$; similarly, for every $\psi \in G(E/K_2)$ we show that $\sigma^{-1}\psi\sigma \in K_1$, so $G(E/K_1)$ and $G(E/K_2)$ are conjugate.

(2) Now we suppose that $G(E/K_1)$ and $G(E/K_2)$ are conjugate subgroups of $G(E/F)$, so there must be some $\sigma \in G(E/F)$ such that

$$\sigma G(E/K_1)\sigma^{-1} = G(E/K_2).$$

We wish to show that σ is an automorphism that maps K_1 onto all of K_2, that is, $\sigma : K_1 \longmapsto\!\!\!\!\twoheadrightarrow K_2$.

Let $\alpha_2 \in K_2$. For every $\varphi \in G(E/K_1)$ we have $\sigma\varphi\sigma^{-1} \in G(E/K_2)$ and therefore

$$\sigma\varphi\sigma^{-1}\alpha_2 = \alpha_2 \qquad \text{or} \qquad \varphi(\sigma^{-1}\alpha_2) = \sigma^{-1}\alpha_2,$$

which shows that $\sigma^{-1}\alpha_2$ is in the fixed field of φ. Since by hypothesis E is normal over F it is also normal over K_1, so K_1 is itself the fixed field of $G(E/K_1)$ and $\sigma^{-1}\alpha_2 \in K_1$. Moreover, every element of K_1 is of the form $\alpha_1 = \sigma^{-1}\alpha_2$ for some $\alpha_2 \in K_2$, and $K_2 = \sigma K_1$, for if $\alpha_1 \in K_1$, then $\sigma\alpha_1 \in K_2$: To see this, we check that $\sigma\alpha_1$ is in the fixed field of $\sigma\varphi\sigma^{-1}$: We have $(\sigma\varphi\sigma^{-1})(\sigma\alpha_1) = \sigma\varphi(\sigma^{-1}\sigma\alpha_1) = \sigma\varphi\alpha_1 = \sigma\alpha_1$. Combining these conclusions, we see that $\sigma : K_1 \longmapsto\!\!\!\!\twoheadrightarrow K_2$. $\quad\|$

EXAMPLE. Let $F = Q$, $E = Q(\omega, \sqrt[3]{2})$, and $K_1 = Q(\sqrt[3]{2})$, $K_2 = Q(\omega\sqrt[3]{2})$, $K_3 = Q(\omega^2\sqrt[3]{3})$. Then $G(E/K_1) = \{\varphi_1, \varphi_3\}$, $G(E/K_2) = \{\underline{\qquad\qquad}\}$, $G(E/K_3) = \{\underline{\qquad\qquad}\}$. These groups are conjugate subgroups of $G(E/F)$. (See Section 2.3 for the definition of $\varphi_1, \ldots, \varphi_6$.)

THEOREM. If K_1 and K_2 are conjugate fields, then $[K_1:F] = [K_2:F]$.

Proof. $[K_2:F] = |G(K_2/F)| = |G(K_1/F)| = [K_1:F]$. $\quad\|$

This theorem also follows very simply from the fact that $K_2 = \sigma K_1$ and automorphism preserve independence relations and therefore degrees. This gives an alternative proof.

THEOREM. Conjugate fields K_1, K_2 are generated by different roots of the same irreducible polynomial $p(x) \in F[x]$.

Proof. Since K_1 is a finite extension of F, it is generated by a single element γ which is the root of some irreducible polynomial $p(x) \in F[x]$, and K_2 must be generated by $\sigma\gamma$, which must then be a root of $\sigma(p(x))$. But all the coefficients of $p(x)$ are in F, so that $\sigma(p(x)) = p(x)$. Both γ and $\sigma\gamma$ are therefore roots of $p(x)$. ‖

EXERCISE. Show that if γ_1 and γ_2 are conjugate in some extension E of Q, then their absolute values are equal: $|\gamma_1| = |\gamma_2|$.

There is a connection between the notion of conjugacy, which relates to other extensions of F within some bigger field E, and that of normalcy, which concerns K and F alone, without embedding K in a larger field. This is expressed more precisely in the definition and theorem which we now state:

DEFINITION. The field K is a *self-conjugate* extension of F if $[K:F]$ is finite and K has no conjugate field other than itself in any finite separable extension E of K.

In other words, the finite extension K is a self-conjugate extension of F if in every extension E of K we have $\sigma(K) = K$, for every $\sigma \in G(E/F)$.

THEOREM

$$K \rhd F \Leftrightarrow K \text{ is a self-conjugate extension of } F.$$

Proof. We already know that

$$K \rhd F \Leftrightarrow K \text{ is the splitting field of a separable polynomial } p(x) \in F[x],$$

so we need only show that
K is the splitting field of a separable polynomial $p(x) \in F[x]$

$$\Leftrightarrow K \text{ is a self-conjugate extension of } F.$$

3.3 CONJUGATE FIELDS AND ELEMENTS

(1) To prove $\Rightarrow$: Suppose K is the splitting field of the separable polynomial $p(x) \in F[x]$ and $K < E$. Then every automorphism $\sigma \in G(E/F)$ maps every coefficient of $p(x)$ into itself; that is, $\sigma(p(x)) = p(x)$. But an automorphism σ maps a root ζ of $p(x)$ into $\sigma(\zeta)$, which must be a root of $\sigma(p(x))$. Therefore, σ merely permutes the roots of $p(x)$ and must map the splitting field of K into itself, which shows that K is self-conjugate.

(2) To prove $\Leftarrow$: We now assume that K is self-conjugate over F and that $E \rhd F$ and $E > K > F$. (There is some such normal extension E of F, because _____.)

We must show that K is the splitting field of some $p(x) \in F[x]$. Since K is finite over F it is a simple extension of F; that is, there is some element $\zeta \in K$ such that $K = F(\zeta)$. Let $\zeta_1, \zeta_2, \ldots, \zeta_n$ be a list of the distinct conjugates of ζ under $G(E/F)$. Every $\zeta_i \in K$, since $\zeta_i = \sigma(\zeta)$ for some $\sigma \in G(E/F)$ and $\sigma(K) = K$. Moreover, σ merely permutes $\zeta_1, \ldots, \zeta_n$. All the coefficients of the polynomial

$$p(x) = (x - \zeta_1) \cdots (x - \zeta_n) = x^n + a_{n-1} x^{n-1} + \cdots + a_0$$

are symmetric in $\zeta_1, \ldots, \zeta_n$ and therefore invariant under every $\sigma \in G(E/F)$. Since we chose E so that $E \rhd F$, this implies that each $a_i \in F$ and $p(x) \in F[x]$. The field K is therefore the splitting field of some $p(x) \in F[x]$. ‖

If we now combine the last theorems, we get a new result that is also already part of the fundamental theorem:

THEOREM.   If $F < K < E$, then

$$F \lhd K \Leftrightarrow G(E/F) \rhd G(E/K).$$

Proof

$F \lhd K \Leftrightarrow K$ is a self-conjugate extension of F

$$\Leftrightarrow \sigma G(E/K) \sigma^{-1} = G(E/K), \qquad \forall \sigma \in G(E/F)$$

$$\Leftrightarrow G(E/F) \rhd G(E/K). ‖$$

3.4   FUNDAMENTAL THEOREM

The first sections of this chapter are intended to familiarize readers with the ideas needed to state the fundamental theorem and to make it appear less formidable, for it really says something quite simple and beautiful and its proof is not at all difficult once the vocabulary and statement of the theorem are clear. After the statement there will therefore first be several specific examples of polynomials, their splitting fields, corresponding groups, and then later the proof will be given.

In its essence the fundamental theorem says that the lattice of subfields of the splitting field E of a given polynomial with coefficients in F looks just like the lattice of subgroups of $G(E/F)$ provided only that inclusion signs are reversed: There is a one-to-one correspondence between the subfields of E and the various subgroups of $G(E/F)$ which reverses inclusions but preserves conjugacy and normality relations and relates the index of subgroup to the degree of the corresponding extension field. We shall write G_K for $G(E/K)$. The field F is often called the ground field.

FUNDAMENTAL THEOREM OF GALOIS THEORY.   Suppose $p(x) \in F[x]$ is separable over F, that E is the splitting field of $p(x)$, and K, K_1, and K_2 are intermediate fields. Then

(1) There is a one-to-one correspondence between subfields K of E and subgroups G of $G(E/F)$ such that

$$K \leftrightarrow G \text{ if and only if } K = F_G$$

Moreover,

$$G_2 < G_1 \text{ if and only if } F_{G_2} > F_{G_1}.$$

(2) K_1 and K_2 are conjugate subfields if and only if G_{K_1} and G_{K_2} are conjugate subgroups.

3.4 FUNDAMENTAL THEOREM

(3) $K \rhd F$ if and only if $G_K \lhd G_F$, and if $F \lhd K$, then $G(K/F) \cong G_F/G_K$.

(4) $[K:F] = \text{ind}_{G_F} G_K$ and $[E:K] = |G_K|$.

EXAMPLES. In all the diagrams that follow, the larger fields are above their subfields, but the larger groups are *below* their subgroups. Connecting lines indicate inclusion relationships, double lines are used if it is normal inclusion (normal subgroup, normal extension field). If $F_1 < F_2$, then the number written along the line from F_1 to F_2 will be $[F_2:F_1]$; if $G_1 > G_2$, then the number along the line from G_1 to G_2 will be $\text{ind}_{G_1} G_2$. An automorphism σ of E over K may be thought of in two different ways: First we think of σ as an automorphism of E which reduces to the identity on K, but we also know from Section 2.3 that corresponding to each σ there is a uniquely defined permutation of the roots of $p(x)$. The abstract group of automorphisms of E, namely, $G(E/F)$, is therefore isomorphic to a specific subgroup of the symmetric group $\mathfrak{S}_n$. (In fact, there are often several different subgroups of $\mathfrak{S}_n$ which are isomorphic to $G(E/F)$. We return to this in Example F.) In all the examples σ_1 will be the identity automorphism and also written as (1).

We shall consider the following polynomials:

A. $x^2 - 5$, ground field Q

B. $x^3 - 5$, ground field Q

C. $x^4 - 5$, ground field Q

D. $(x^2 - 2)(x^3 - 5)$, ground field Q

E. $(x^2 - 2)(x^2 - 5)(x^2 - 7)$, ground field Q

F. $x^6 - 6x^4 - 10x^3 + 12x^2 - 60x + 17$, ground field Q

G. $x^4 - 5$, ground field $Q(i)$

H. $x^3 + x + 1$, ground field $GF(2)$

EXAMPLE A. Here

$$p(x) = x^2 - 5,$$
$$F = Q,$$
$$E = Q(\sqrt{5}).$$

Let the roots of $p(x)$ be

$$\alpha_1 = \sqrt{5},$$

$$\alpha_2 = -\sqrt{5}.$$

So the automorphisms of E are determined by

$$\sigma_1 : \begin{cases} \alpha_1 \to \alpha_1 \\ \alpha_2 \to \alpha_2 \end{cases} \qquad \sigma_2 : \begin{cases} \alpha_1 \to \alpha_2 \\ \alpha_2 \to \alpha_1 \end{cases}$$

The corresponding permutations are

$$\sigma_1 \text{ corresponds to } (1),$$

$$\sigma_2 \text{ corresponds to } (12).$$

The group multiplication table is

Second factor / First factor	σ_1	σ_2
σ_1	σ_1	σ_2
σ_2	σ_2	σ_1

or

Second factor / First factor	(1)	(12)
(1)	(1)	(12)
(12)	(12)	(1)

We have $G(E/F) = \{\sigma_1, \sigma_2\} = \{(1), (12)\} = C_2$, and $[E:F] = 2 = |G(E/F)|$. The diagram is very simple:

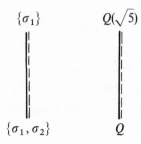

3.4 FUNDAMENTAL THEOREM

EXAMPLE B. Here

$$p(x) = x^3 - 5$$

$$F = Q,$$

$$E = Q(\omega, \sqrt[3]{5}).$$

A basis for E as a vector space over F consists of the _____ elements _____, and therefore $\dim_F E = [E:F] =$ _____. Let $\alpha_1 = \sqrt[3]{5}$, $\alpha_2 = \omega\sqrt[3]{5}$, $\alpha_3 = \omega^2\sqrt[3]{5}$. The general element of E is of the form _____. We can list the 6 automorphisms of E as follows:

The automorphism	σ_1	σ_2	σ_3	σ_4	σ_5	σ_6
sends ω into	ω	ω	ω	ω^2	ω^2	ω^2
and $\alpha_1 = \sqrt[3]{5}$ into	α_1	α_2	α_3	α_1	α_2	α_3
Therefore, it sends $\alpha_2 = \omega\sqrt[3]{5}$ into	α_2	α_3				
and $\alpha_3 = \omega^2\sqrt[3]{5}$ into	α_3	α_1				
It can therefore be represented by the permutation	(1)	(123)				(13)

Therefore, $G_F = G(E/F) = \{\sigma_1, \ldots, \sigma_6\} = \{(1),$ _____ $\} = \mathfrak{S}_3,$ and we have $|G_F| =$ _____. The group multiplication is described by the next table.

then →	σ_1	σ_2	σ_3	σ_4	σ_5	σ_6
First apply ↘ σ_1						
σ_2						
σ_3						
σ_4						
σ_5						
σ_6						

This group is not commutative, so it matters very much which automorphism is applied first. For example, if $a = (\omega + \sqrt[3]{5})$, then

$$a \overset{\sigma_3}{\rightarrow} \underline{\hspace{4cm}} \overset{\sigma_4}{\rightarrow} \underline{\hspace{4cm}},$$

while

$$a \overset{\sigma_4}{\rightarrow} \underline{\hspace{4cm}} \overset{\sigma_3}{\rightarrow} \underline{\hspace{4cm}}.$$

This may be written as $a(\sigma_3\sigma_4) = \omega^2 + \omega\sqrt[3]{5}$ and $a(\sigma_4\sigma_3) = \underline{\hspace{3cm}}$, with the understanding that $a(\sigma_3\sigma_4) = (a\sigma_3)\sigma_4$, but often the older notation is used and one writes $(\sigma_4\sigma_3)(a) = \omega^2 + \omega\sqrt[3]{5}$ and $(\sigma_3\sigma_4)(a)\underline{\hspace{3cm}}$, where $(\sigma_4\sigma_3)(a) = \sigma_4(\sigma_3(a))$. We shall always try to make sure that the notation is clarified by the immediate context and insert parentheses in any case of ambiguity. In any reference book it is necessary to check carefully which notation is being used.

3.4 FUNDAMENTAL THEOREM

Next we list all the subgroups of $\mathfrak{S}_3$ by listing the permutations contained in each. They are

$$\mathfrak{S}_3 = \{\underline{\hspace{6cm}}\},$$

$$\mathfrak{A}_3 = \{\underline{\hspace{5cm}}\},$$

$$H_1 = \{\underline{\hspace{4cm}}\},$$

$$H_2 = \{\underline{\hspace{4cm}}\},$$

$$H_3 = \{\underline{\hspace{4cm}}\},$$

$$I = (1).$$

Of these $\mathfrak{A}_3 \triangleleft \mathfrak{S}_3$, because $\underline{\hspace{7cm}}$ and H_1, H_2, H_3 are conjugate, because $\underline{\hspace{6cm}}$.

If we arrange these groups in the form of an inclusion lattice with the smaller on top we get Figure 3.3(a). Figure 3.3(b) is the lattice of subfields of the splitting field

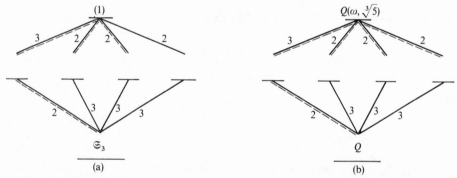

FIGURE 3.3

$E = Q(\omega, \sqrt[3]{5})$ over the base field $F = Q$. If $F < K < E$, then $[E:K] \cdot [K:F] = [E:F] = 6$, so that we must have $[K:F] = 2$ or 3 for any $K \neq E, F$. The following considerations help us put the right fields in the right spaces:

Case 1. If $[K:F] = 2$, then $\sqrt[3]{5} \notin K$, because $\underline{\hspace{5cm}}$

_____ and in fact of the various basis elements of K as a vector space over F, only _____ can be in K. So in this case $K =$ _____.

Case 2. Suppose $[K:F] = 3$. Then K is three dimensional as a vector space over F. If 1, ω, and any other basis element are in K, we already have $K = E$, so we know that $\omega \notin K$. The basis elements contained in K are therefore $1,$ _____, _____, or $1,$ _____, _____, or $1,$ _____, _____, so that the three choices for K are $F_1 = Q(\underline{\hspace{1cm}})$, $F_2 = Q(\underline{\hspace{1cm}})$, and $F_3 = Q(\underline{\hspace{1cm}})$. We see, therefore, that in this particular case there are just as many intermediate fields as there are intermediate subgroups of $\mathfrak{S}_3$.

Three of the subgroups are conjugate, and so are the three fields F_1, F_2, F_3. For example the automorphism _____ will transform F_1 into F_2, _____ will transform F_2 into F_3, and so on. With just a little care, one can arrange F_1, F_2, F_3 in part (b) in such a way that each entry is the fixed field of the group H_i in the corresponding position on part (a) and also rearrange the numbering of the subscripts so that F_i is the fixed field of H_i. Once the diagrams are completed, it is easy to check that all the conclusions of the fundamental theorem hold in this case.

These two cases illustrate the general situation, but they also leave open a few more questions: What about equations whose group is not all of the symmetric group? Do all diagrams of field inclusions exhibit a certain symmetry as in these examples? What happens when the ground field is not Q but a bigger field? When char $F \neq 0$? Sometimes the examination of special examples suggests more detailed theorems about equations and their groups, so we continue with

EXAMPLE C

$$p(x) = x^4 - 5,$$

$$F = Q,$$

$$E = Q(i, \sqrt[4]{5}).$$

3.4 FUNDAMENTAL THEOREM

Notice that i is a primitive fourth root of unity, as i, i^2, i^3 are all different from 1, and $i^4 = 1$. (Of course, here, as always, $i = \sqrt{-1}$.) A basis for E over F is formed by the 8 elements $1, i, 5^{1/4}, i5^{1/4}, 5^{1/2}, i5^{1/2}, 5^{3/4}, i5^{3/4}$. Let $\alpha_1 = 5^{1/4}, \alpha_2 = i5^{1/4}, \alpha_3 = -5^{1/4}$ $\alpha_4 = -i5^{1/4}$ be the roots of $p(x)$.

We list the 8 automorphisms of E:

The automorphism	σ_1	σ_2	σ_3	σ_4	σ_5	σ_6	σ_7	σ_8
sends i into	i	i	i	i	$-i$	$-i$	$-i$	$-i$
and $\alpha_1 = 5^{1/4}$ into	α_1	α_2	α_3	α_4	α_1	α_2	α_3	α_4
Therefore, it sends $\alpha_2 = i5^{1/4}$ into	α_2	α_3				α_1		
$\alpha_3 = -5^{1/4}$ into	α_3	α_4				α_4		
$\alpha_4 = -i5^{1/4}$ into	α_4	α_1				α_3		
σ_i can therefore be represented by the permutation	(1)	(1234)				(12)(34)		

For example, the computations for σ_6 are

$$\sigma_6(\alpha_2) = \sigma_6(i5^{1/4}) = \sigma_6(i)\sigma_6(5^{1/4}) = (-i)(-5^{1/4}) = 5^{1/4} = \alpha_1,$$

$$\sigma_6(\alpha_3) = \sigma_6(-5^{1/4}) = -\alpha_2 = \alpha_4,$$

$$\sigma_6(\alpha_4) = \sigma_6(-i5^{1/4}) = -\alpha_1 = \alpha_3.$$

Therefore, in this example we have $G_F = G(E/F) = \{(1),\ (1234),$ _____, _____, _____, $(12)(34),$ _____, _____$\}$, for which we have the following group multiplication table:

First apply ↘ / then →	σ_1	σ_2	σ_3	σ_4	σ_5	σ_6	σ_7	σ_8
σ_1	σ_1	σ_2	σ_3	σ_4	σ_5	σ_6	σ_7	σ_8
σ_2	σ_2							
σ_3	σ_3							
σ_4	σ_4							
σ_5	σ_5							
σ_6	σ_6				σ_1			
σ_7	σ_7							
σ_8	σ_8							

It is evident from the table that the group G_F in this example is non-Abelian and contains 8 elements. There are only two essentially different (non-isomorphic) non-Abelian groups of order 8: the quaternion group, which contains 1 element of order 2, and the dihedral group, which contains 5 elements of order 2. The group here is therefore the _____ group.

3.4 FUNDAMENTAL THEOREM

We again want to list all subgroups H of G_E, but this time we shall list next to each subgroup H its fixed field F_H. For example, the fixed field of the subgroup $\{(1), \sigma_6\}$ consists of all elements $x \in E$ for which $\sigma_6(x) = x$. If

$$x = a_1 + ia_2 + 5^{1/4}a_3 + i5^{1/4}a_4 + 5^{1/2}a_5 + i5^{1/2}a_6 + 5^{3/4}a_7 + i5^{3/4}a_8,$$

then

$$\sigma_6(x) = a_1 + \underline{\hspace{8cm}}.$$

If, therefore, $x = \sigma_6(x)$, we must have $a_2 = a_5 = 0$, $a_3 = a_4$, and $a_7 = -a_8$, and $F_H = \{a_1 + 5^{1/4}(1 + i)a_3 + i5^{1/2}a_6 + 5^{3/4}(1 - i)a_7 | a_i \in Q\}$. It is perhaps not immediately clear on inspection that the elements of F_H are indeed closed under multiplication and reciprocal, so one should check it. Do we have $F_H = Q(5^{1/4}(1 + i))$? $\underline{\hspace{2.5cm}}$ We also see that $[F_H : Q] = \underline{\hspace{2.5cm}}$, and $\text{ind}_{G_F} H = \underline{\hspace{2.5cm}}$. So in order to find all subfields of E, we first list all $\sigma_i x$. (If there is no ambiguity, parentheses around x are often omitted.)

$$x = \sigma_1 x = a_1 + a_2 i + 5^{1/4}a_3 + i5^{1/4}a_4 + 5^{1/2}a_5 + i5^{1/2}a_6 + 5^{3/4}a_7 + i5^{3/4}a_8,$$

$$\sigma_2 x = \underline{\hspace{9cm}},$$

$$\sigma_3 x = \underline{\hspace{9cm}},$$

$$\sigma_4 x = \underline{\hspace{9cm}},$$

$$\sigma_5 x = \underline{\hspace{9cm}},$$

$$\sigma_6 x = \underline{\hspace{9cm}},$$

$$\sigma_7 x = \underline{\hspace{9cm}},$$

$$\sigma_8 x = \underline{\hspace{9cm}},$$

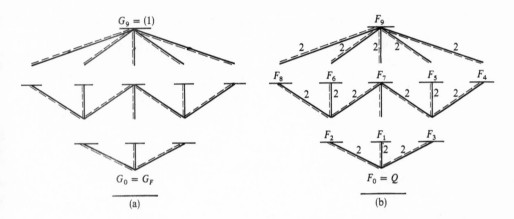

FIGURE 3.4

The subgroups of G_F are	The corresponding subfields are

$G_0 = G_F = \{(1), (1234), (13)(24), (1432),$ $F_0 = Q,$
$\quad (24), (12)(34), (13), (14)(23)\},$

$G_1 = \{(1), (1234), (13)(24), (1432)\},$ $\quad F_1 = Q(i),$

$G_2 = \{(1), (12)(34), (13)(24), (14)(23)\},$ $\quad F_2 = Q(i5^{1/2}),$

$G_3 = \{(1), (13), (24), (13)(24)\},$ $\quad F_3 = Q(5^{1/2}),$

$G_4 = \{(1), (13)\},$ $\quad F_4 = Q(i5^{1/4}),$

$G_5 = \{(1), (24)\},$ $\quad F_5 = Q(5^{1/4}),$

$G_6 = \{(1), (12)(34)\},$ $\quad F_6 = Q(5^{1/4}(1 + i)),$

$G_7 = \{(1), (13)(24)\},$ $\quad F_7 = Q(i, 5^{1/2}),$

$G_8 = \{(1), (14)(23)\},$ $\quad F_8 = Q(5^{1/4}(1 - i)),$

$G_9 = \{(1)\}.$ $\quad F_9 = E = Q(i, 5^{1/4}).$

3.4 FUNDAMENTAL THEOREM

Suppose we wish to check that F_2 is really the fixed field of G_2. One way to do this is the following: We have

$$i5^{1/2} = \alpha_1\alpha_2 = (1)\alpha_1\alpha_2 = (12)(34)\alpha_1\alpha_2$$

and

$$(13)(24)\alpha_1\alpha_2 = \alpha_3\alpha_4 = (-5^{1/4})(-i5^{1/4}) = \alpha_1\alpha_2,$$

$$(14)(23)\alpha_1\alpha_2 = \underline{\hspace{3cm}} = \alpha_1\alpha_2.$$

Therefore, $i5^{1/2}$ is fixed by G_2 and $F_2 \subseteq F_{G_2}$. Checking the table listing $\sigma_i x$ ($i = 1, \ldots, 8$) shows that we have $F_2 = F_{G_2}$. The other fields can all be checked in a similar manner. The groups and fields are arranged as lattices according to Figure 3.4.

Notice that although F_8 is a normal extension of F_2, it is not a normal extension of F_0. In fact, F_8 and F_6 are conjugate over F_0, since $F_6 = \sigma_5(F_8)$, and F_5 and F_4 are also conjugate, since $F_5 = \underline{\hspace{1.5cm}} (F_4)$. The field F_7 is a normal extension of F_0, and we should expect $G_7 \triangleleft G_0$: If we let π be any permutation of G_0, this implies checking that $\pi^{-1}(13)(24)\pi \in G_7$. Carrying this out for $\pi = (1234)$ we get $\pi^{-1} = (1432)$, and $\pi^{-1}(13)(24)\pi = \underline{\hspace{5cm}} \in G_7$. Moreover, $G(F_7/F_0)$ should be isomorphic to G_7/G_0: The group $G(F_7/F_0)$ is the group formed by the 4 automorphisms of F_7 which leave Q fixed:

$$\sigma_1 : \begin{cases} i \to i \\ \sqrt{5} \to \sqrt{5} \end{cases} \qquad \sigma_3 : \begin{cases} i \to -i \\ \sqrt{5} \to \sqrt{5} \end{cases}$$

$$\sigma_2 : \begin{cases} i \to i \\ \sqrt{5} \to -\sqrt{5} \end{cases} \qquad \sigma_4 : \begin{cases} i \to -i \\ \sqrt{5} \to -\sqrt{5} \end{cases}$$

Each of these automorphisms is of period 2, so $G(F_7/F_0) \cong \mathfrak{B}$. Clearly, $|G_0/G_7| = |G_0|/|G_7| = \frac{8}{2} = 4$, but we must still check that each element of G_0/G_7 is of order 2 in order to have $G(F_7/F_0) \cong G_0/G_7$. We leave this to the reader as an exercise in group theory.

EXAMPLE D. Let

$$p(x) = (x^2 - 2)(x^3 - 5),$$

$$F = Q,$$

$$E = Q(\sqrt{2}, \sqrt[3]{5}, \omega).$$

The roots of $p(x)$ are

$$\alpha_1 = \sqrt[3]{5},$$

$$\alpha_2 = \omega\sqrt[3]{5},$$

$$\alpha_3 = \omega^2\sqrt[3]{5},$$

$$\alpha_4 = \sqrt{2},$$

$$\alpha_5 = -\sqrt{2}.$$

There are 12 automorphisms, because $[E:F] = 12$:

The automorphism	σ_1	σ_2	σ_3	σ_4	σ_5	σ_6	σ_7	σ_8	σ_9	σ_{10}	σ_{11}	σ_{12}
is determined by $\sqrt[3]{5} \rightarrow$	α_1	α_2	α_3	α_1	α_2	α_3	α_1	α_2	α_3	α_1	α_2	α_3
$\omega \rightarrow$	$\omega \longrightarrow$			$\omega^2 \longrightarrow$			$\omega \longrightarrow$			$\omega^2 \longrightarrow$		
$\sqrt{2} \longrightarrow$	$\sqrt{2} \longrightarrow$						$-\sqrt{2} \longrightarrow$					
and corresponds to the permutation	(1)	(123)					(45)	(123)(45)				

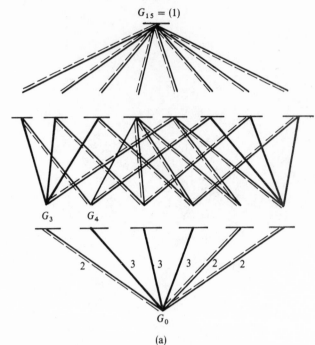

$G_{15} = (1)$

G_3 G_4

2   3   3   3   2   2

G_0

(a)

FIGURE 3.5 $p(x) = (x^2 - 2)(x^3 - 5)$

$G_0 = G_F = \{\underline{\hspace{8cm}}\}$,

$G_1 = \{\underline{\hspace{3cm}}\}$

$G_2 = \{\underline{\hspace{3cm}}\}$,

$G_3 = \{\underline{\hspace{3cm}}\}$,

$G_4 = \{\underline{\hspace{3cm}}\}$,

$G_5 = \{\underline{\hspace{3cm}}\}$,

$G_6 = \{\underline{\hspace{3cm}}\}$,

$G_7 = \{\underline{\hspace{3cm}}\}$,

$G_8 = \{\underline{\hspace{3cm}}\}$,

$G_9 = \{\underline{\hspace{3cm}}\}$,

$G_{10} = \{\underline{\hspace{5cm}}\}$,

$G_{11} = \{\underline{\hspace{5cm}}\}$,

$G_{12} = \{\underline{\hspace{5cm}}\}$,

$G_{13} = \{\underline{\hspace{5cm}}\}$,

$G_{14} = \{\underline{\hspace{5cm}}\}$,

$G_{15} = \{(1)\}$.

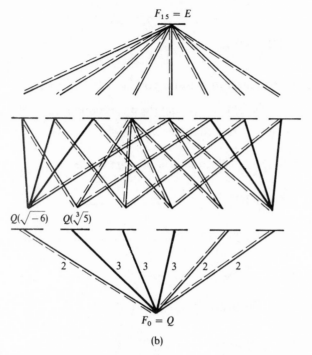

$F_{15} = E$

$Q(\sqrt{-6})$ $Q(\sqrt[3]{5})$

2   3   3   3   2   2

$F_0 = Q$

(b)

FIGURE 3.5 (*continued*)

$$F_0 = Q,$$
$$F_1 = Q(\sqrt{2}),$$
$$F_2 = Q(\omega),$$
$$F_3 = Q(\sqrt{-6}),$$
$$F_4 = Q(\sqrt[3]{5}),$$
$$F_5 = Q(\omega\sqrt[3]{5}),$$
$$F_6 = Q(\omega^2\sqrt[3]{5}),$$
$$F_7 = Q(\omega, \sqrt{2}),$$
$$F_8 = Q(\sqrt{-6}, \sqrt[3]{5}),$$
$$F_9 = Q(\sqrt{-6}, \omega\sqrt[3]{5}),$$
$$F_{10} = Q(\sqrt{-6}, \omega^2\sqrt[3]{5}),$$
$$F_{11} = Q(\omega, \sqrt[3]{5}),$$
$$F_{12} = Q(\sqrt{2}, \sqrt[3]{5},$$
$$F_{13} = Q(\sqrt{2}, \omega\sqrt[3]{5},$$
$$F_{14} = Q(\sqrt{2}, \omega^2\sqrt[3]{5}),$$
$$F_{15} = Q(\sqrt{2}, \sqrt[3]{5}, \omega).$$

3.4 FUNDAMENTAL THEOREM

Since the first 6 of these form a subgroup isomorphic to $\mathfrak{S}_3$, it is easily seen that $G(E/F) \cong \mathfrak{S}_3 \times C_2$. The subgroups and the corresponding subfields are listed below Figure 3.5. They can be arranged as shown there.

We have three sets of fields that are conjugate over Q:

$$\text{set } 1 = \{Q(\sqrt[3]{5}), \underline{\hspace{2cm}}, \underline{\hspace{2cm}}\},$$

$$\text{set } 2 = \{Q(\sqrt{-6}, \sqrt[3]{5}), \underline{\hspace{2cm}}, \underline{\hspace{2cm}}\},$$

$$\text{set } 3 = \{Q(\sqrt{2}, \sqrt[3]{5}), \underline{\hspace{2cm}}, \underline{\hspace{2cm}}\}$$

and seven self-conjugate fields: Q, $Q(\sqrt{-6})$, $Q(\omega)$, $\underline{\hspace{2cm}}$, $\underline{\hspace{2cm}}$, $\underline{\hspace{2cm}}$, and of course $E = Q(\omega, \sqrt{2}, \sqrt[3]{5})$.

EXAMPLE E.   Let

$$p(x) = (x^2 - 2)(x^2 - 5)(x^2 - 7),$$

$$F = Q,$$

$$E = Q(\sqrt{2}, \sqrt{5}, \sqrt{7}).$$

The roots of $p(x)$ are $\alpha_1 = \sqrt{2}$, $\alpha_2 = -\sqrt{2}$, $\alpha_3 = \sqrt{5}$, $\alpha_4 = -\sqrt{5}$, $\alpha_5 = \sqrt{7}$, $\alpha_6 = -\sqrt{7}$. We have $[E:F] = 8$, so $|G_F| = 8$ and there are 8 automorphisms, each of order 2. It is therefore easy to check that $G_F \cong C_2 \times C_2 \times C_2$, whose diagram is given in Figure 3.6 together with a list of subfields.

The automorphisms are

The automorphism	σ_1	σ_2	σ_3	σ_4	σ_5	σ_6	σ_7	σ_8
sends	$\sqrt{2}\to\ \sqrt{2}$	$-\sqrt{2}$	$\sqrt{2}$	$-\sqrt{2}$	$\sqrt{2}$	$-\sqrt{2}$	$\sqrt{2}$	$-\sqrt{2}$
	$\sqrt{5}\to\ \sqrt{5}\longrightarrow$		$-\sqrt{5}\longrightarrow$		$\sqrt{5}\longrightarrow$		$-\sqrt{5}\longrightarrow$	
	$\sqrt{7}\to\ \sqrt{7}\longrightarrow$				$-\sqrt{7}\longrightarrow$			

This time, for a change, we list the automorphisms of G_F as elements of the subgroups of G_F rather than the corresponding permutations.

Make sure that the numbering of the groups corresponds to the numbering of the fields!

In this example the Galois group is Abelian, so all subgroups are invariant and all subfields are self-conjugate.

This example brings up the next obvious question. Which is easier: making a list of all the subgroups of $G(E/F)$ or all the subfields of E? In general, it turns out to be far easier to list all the subgroups first. In listing subfields it is very easy both to overlook some of them and also to list some of them twice without realizing it. For example, $Q(\sqrt{10}, \sqrt{14})$ is easily overlooked, and what about $Q(\sqrt{10}, \sqrt{35})$, $Q(\sqrt{10} + \sqrt{35})$, $Q(\sqrt{10} + \sqrt{70})$, and so on? Where do they fit in? You might try to cover up the list of subfields of $Q(\sqrt{2}, \sqrt{5}, \sqrt{7})$ and try to reconstruct it. This shows that the fundamental theorem can also be used to discover all the subfields of a given field without forgetting or duplicating any and that this is not always an easy task.

3.4 FUNDAMENTAL THEOREM

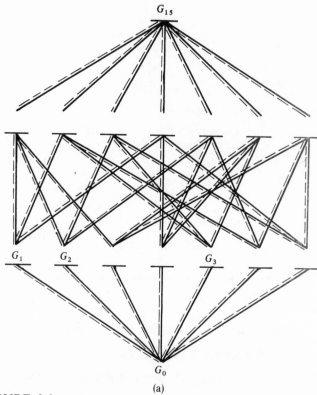

FIGURE 3.6

$G_0 = \{\sigma_1, \ldots, \sigma_8\}$,

$G_1 = \{\sigma_1, \sigma_3, \sigma_5, \sigma_7\}$,

$G_2 = \{\sigma_1, \sigma_2, \sigma_5, \sigma_6\}$,

$G_3 = \{$_____$\}$,

$G_4 = \{$_____$\}$,

$G_5 = \{$_____$\}$,

$G_6 = \{$_____$\}$,

$G_7 = \{$_____$\}$,

$G_8 = \{$_____$\}$,

$G_9 = \{$_____$\}$,

$G_{10} = \{$_____$\}$,

$G_{11} = \{$_____$\}$,

$G_{12} = \{$_____$\}$,

$G_{13} = \{$_____$\}$,

$G_{14} = \{$_____$\}$,

$G_{15} = \{\sigma_1\}$.

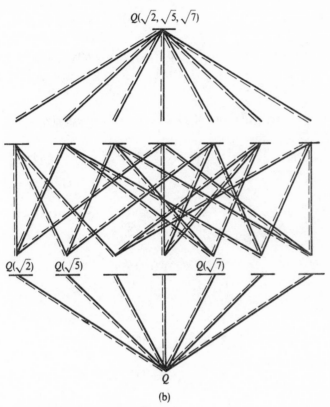

(b)

FIGURE 3.6 (*continued*)

$F_0 = Q,$

$F_1 = Q(\sqrt{2}),$

$F_2 = Q(\sqrt{5}),$

$F_3 = Q(\sqrt{7}),$

$F_4 = Q(\sqrt{10}),$

$F_5 = Q(\sqrt{14}),$

$F_6 = Q(\sqrt{35}),$

$F_7 = Q(\sqrt{70}),$

$F_8 = Q(\sqrt{2}, \sqrt{5}),$

$F_9 = Q(\sqrt{2}, \sqrt{7}),$

$F_{10} = Q(\sqrt{5}, \sqrt{7}),$

$F_{11} = Q(\sqrt{2}, \sqrt{35}),$

$F_{12} = Q(\sqrt{5}, \sqrt{14}),$

$F_{13} = Q(\sqrt{7}, \sqrt{10}),$

$F_{14} = Q(\sqrt{10}, \sqrt{14}),$

$F_{15} = Q(\sqrt{2}, \sqrt{5}, \sqrt{7}).$

3.4 FUNDAMENTAL THEOREM

EXAMPLE F. Let

$$p(x) = x^6 - 6x^4 - 10x^3 + 12x^2 - 60x + 17,$$

$$F = Q,$$

$$E = Q(\alpha_1, \alpha_2, \alpha_3, \alpha_4, \alpha_5, \alpha_6), \qquad \text{where } p(\alpha_i) = 0.$$

The roots of $p(x)$ are

$$\alpha_1 = \sqrt{2} + \sqrt[3]{5},$$

$$\alpha_2 = \sqrt{2} + \omega\sqrt[3]{5},$$

$$\alpha_3 = \sqrt{2} + \omega^2\sqrt[3]{5},$$

$$\alpha_4 = -\sqrt{2} + \sqrt[3]{5},$$

$$\alpha_5 = -\sqrt{2} + \omega\sqrt[3]{5},$$

$$\alpha_6 = -\sqrt{2} + \omega^2\sqrt[3]{5},$$

as can easily be checked. Therefore, the splitting field is just the same as in Example D: Clearly each α_i is a combination of $\sqrt{2}$, $\sqrt[3]{5}$, ω, and conversely each of these can be expressed in terms of $\alpha_1, \ldots, \alpha_6$: We have

$$\sqrt{2} = \tfrac{1}{2}(\alpha_1 - \alpha_4),$$

$$\sqrt[3]{5} = \underline{\hspace{3cm}},$$

$$\omega = \frac{\alpha_2 - \sqrt{2}}{\alpha_1 - \sqrt{2}} = \underline{\hspace{4cm}}.$$

As an abstract group, $G(E/F)$ is therefore the same in this example as in Example D, but as a permutation group it is not: In D we found a subgroup of $\mathfrak{S}_5$ which was isomorphic to $G(E/F)$; here we shall get a transitive subgroup of $\mathfrak{S}_6$.

Using the same $\sigma_1, \ldots, \sigma_{12}$ as in Example D we get the following table:

The automorphism											
σ_1	σ_2	σ_3	σ_4	σ_5	σ_6	σ_7	σ_8	σ_9	σ_{10}	σ_{11}	σ_{12}
now corresponds to the permutation											
(1)	(123)(456)					(14)(25)(36)					
while in Example D it corresponded to											
(1)	(123)					(45)	(123)(45)				

Since $\mathfrak{S}_5 < \mathfrak{S}_6$, both the permutation group obtained in Example D and the one obtained here are subgroups of $\mathfrak{S}_6$ and both are isomorphic to $G(E/F)$. In Chapter IV we shall see that the Galois group of a polynomial will actually determine which polynomials are solvable in radicals and even how to carry out the solution of solvable polynomials, but for this last step we shall need the Galois group expressed as a permutation group on the roots of $p(x)$; the abstract group alone is not enough. This is a reasonable requirement, for you clearly do not need to find the roots of $(x^2 - 2)(x^3 - 5)$ by the more elaborate methods necessary for the irreducible polynomial $x^6 - 6x^4 - 10x^3 + 12x^2 - 60x + 17$.

EXAMPLE G.   Let

$$p(x) = x^4 - 5,$$

$$F = Q(i),$$

$$E = Q(i, \sqrt[4]{5}) = F(\sqrt[4]{5}).$$

3.4 FUNDAMENTAL THEOREM

This is almost the same as Example C, but now $Q(i)$ is the fixed field, so the only automorphisms of E that are in the Galois group are $\sigma_1, \sigma_2, \sigma_3, \sigma_4$ (in the notation of Example C). The group $G(E/F)$ is therefore cyclic of order 4. (If p is a prime and F contains all nth roots of unity, is $F(\sqrt[n]{p})$ always cyclic? [A, p. 61].)

The diagram for this group is a sublattice of the one in Example C and is given in Figure 3.7. In all the other examples, taking some field K other than Q as the ground field would also just have reduced the field lattice by simply leaving all fields containing K and erasing all others.

G_9

G_1

(a)

(b)

FIGURE 3.7

EXAMPLE H.   Let

$$p(x) = x^3 + x + 1,$$

$$F = GF(2) = Z_2 = \{0, 1\}_{\text{mod } 2},$$

$$E = F(\alpha_1, \alpha_2, \alpha_3).$$

In this example we have a field whose characteristic is not 0. In a field of characteristic 0, no irreducible polynomial has repeated roots, and so every polynomial is separable. We now must check not only whether $p(x)$ is irreducible, but also whether it is separable. If $p(x)$ were reducible, it would have at least one linear factor $(x - a)$ with $a \in F$. Then $p(a)$ would have to be 0. However, $p(0) = $ _____, and $p(1) = $ _____, so $p(x)$ has no roots in F, and therefore no linear factor. So $p(x)$ is irreducible. The theorems quoted at the end of Section 3.2 assure us that $p(x)$ is separable, but this can easily be checked independently: We know that $\alpha_1, \alpha_2, \alpha_3$ are not zero, as $p(0) \neq 0$. So suppose that $\alpha_1 = \alpha_2 = \alpha_3$. Then, working mod 2, we get $p(x) = (x - \alpha_1)^3 = x^3 - 3\alpha_1 x^2 + 3\alpha_1^2 x - \alpha_1^3 = x^3 + \alpha_1 x^2 + \alpha_1^2 x + \alpha_1^3 = x^3 + x + 1$. Therefore, $\alpha_1 x^2 = 0$, so $\alpha_1 = 0$. This is impossible, so $p(x)$ cannot have three equal roots. Suppose now that $\alpha_1 = \alpha_2 \neq \alpha_3$. Then, of course again mod 2, we get $p(x) = (x - \alpha_1)^2(x - \alpha_3) = (x^2 + \alpha_1^2)(x - \alpha_3) = x^3 - \alpha_3 x^2 + \alpha_1^2 x - \alpha_1^2 \alpha_3$, so again $\alpha_3 = 0$, which is impossible. Therefore, $p(x)$ is separable.

If we work in $F(\alpha_1)$, then $(x - \alpha_1)$ must be a factor of $p(x)$. In fact, we get

$$x^3 + x + 1 = (x - \alpha_1)(x^2 + \alpha_1 x + (1 + \alpha_1^2))$$

$$= (x - \alpha_1)(x - \alpha_1^2)(x - (\alpha_1^2 + \alpha_1)).$$

In checking this, remember that $\alpha_1^3 + \alpha_1 + 1 = 0$ and we are working mod 2. (Minus signs are therefore not really necessary.) This factorization shows that $F(\alpha_1) = F(\alpha_1, \alpha_2, \alpha_3) = E$ and that $\alpha_2 = \alpha_1^2$, $\alpha_3 = \alpha_1^2 + \alpha_1$. Therefore, $[E:F] = 3$, so $|G(E/F)| = 3$, and the only possibility is $G(E/F) = C_3$.

As a permutation group, $G(E/F) = \{(1), (123), (132)\}$. Had we proved the following theorem, we could have foreseen this conclusion without factoring $p(x)$.

THEOREM. The Galois group of any finite field is cyclic. Moreover, if $E = GF(p^n)$ and $F = GF(p)$, then $G(E/F) \cong C_n$ [Ad, p. 129; vdW, p. 117 or Sec. 4.11.]

Many of the calculations necessary in the construction of the preceding examples

3.4 FUNDAMENTAL THEOREM

are longer and more intricate than those involved in the proof of the fundamental theorem, which is our next order of business.

Before starting to read the proof, you should now reread the statement of the theorem and remember that we write G_K for $G(E/K)$ and F_G for the field consisting of all those elements of E which are left fixed by *every* $\sigma \in G$. Therefore, we always have $K < F_{G_K}$, but the automorphisms of G_K might conceivably leave fixed a field larger than just K. In fact, this always happens unless E is a normal extension of K, for the definition of normality (Section 3.2) says that $E \vartriangleright K$ if and only if $F_{G_K} = K$.

Proof of the Fundamental Theorem. In reading this proof, remember that if E is a normal extension of F, then E is also a normal extension of any intermediate field K. This is used at several points in the proof, usually to conclude that $F_{G_K} = K$.

Proof of Part (1). We define a mapping T from the set $\{K\}$ of subfields of E to the set $\{G\}$ of subgroups of G_F which will establish the required one-to-one correspondence by setting $T(K) = G_K$. We must show that

(1a) T is a well defined and one to one.

(1b) T is onto the entire set of subgroups of G_F.

(1c) $T(K) = G \Leftrightarrow K = F_G$.

(1d) T reverses inclusions.

(1a) T is well defined and one to one: For each field $K < E$ the group G_K is uniquely determined by the definition $G_K = \{\sigma \mid \sigma \in G(E/F) \text{ and } \sigma \restriction K = I\}$, so $T(K)$ is well defined. To prove that T is one to one, we must show that

$$T(K_1) = T(K_2) \Rightarrow K_1 = K_2.$$

We have

$$T(K_1) = T(K_2) \Rightarrow G_{K_1} = G_{K_2}, \qquad \text{by the definition of } T,$$

$$\Rightarrow F_{G_{K_1}} = F_{G_{K_2}},$$

$$\Rightarrow K_1 = K_2, \qquad \text{since } E \vartriangleright K.$$

(1b) T is onto: Let H be any subgroup of G_F. We must show that there is some field $K > F$ for which $T(K) = H$. We shall show that F_H is the required field K—namely that $T(F_H) = H$: This is true since $T(F_H) = G_{F_H}$, by definition of T, and

$$F_H = F_{G_{F_H}}, \qquad \text{by the normality of } E \text{ over } F_H.$$

Corollary 2 of Section 2.4 (p. 67) says that

$$F_H = F_{G_{F_H}} \Rightarrow H = G_{F_H},$$

which is exactly what we wanted.

(1c) $T(K) = H \Leftrightarrow K = F_H$: To prove $\Rightarrow$ we have

$$T(K) = H \Rightarrow G_K = H, \qquad \text{by the definition of } T,$$

$$\Rightarrow F_{G_K} = F_H,$$

$$\Rightarrow K = F_{G_K} = F_H, \qquad \text{since } E \rhd K.$$

To prove $\Leftarrow$ we have

$$K = F_H \Rightarrow K = F_{G_K} = F_H, \qquad \text{since } E \rhd K,$$

$$\Rightarrow G_K = H, \qquad \text{by Corollary 2 of Section 2.4,}$$

$$\Rightarrow T(K) = H, \qquad \text{by the definition of } T.$$

[This part actually proves that if we define the function $S(H)$ by setting $S(H) = F_H$, then $S = T^{-1}$ and $T = S^{-1}$. Since a function has an inverse if and only if it is one to one and onto, we could have deduced (1a) from (1c). As we saw, however, it is easy to prove separately.]

(1d) $H_1 < H_2 \Rightarrow F_{H_1} > F_{H_2}$: This follows immediately from the definition of the fixed field.

This proves the first part of the theorem.

3.4 FUNDAMENTAL THEOREM

Proof of Part (2). This is merely a restatement of the first theorem of Section 3.3 (p. 89).

Proof of Part (3). The first statement of this part was already proved at the end of Section 3.3. To prove that $G(K/F) \cong G_F/G_K$ we define a function ψ which maps $G(K/F)$ into G_F/G_K and then prove that it is an isomorphism. For every $\sigma \in G(K/F)$, we define

$$\psi(\sigma) = [\bar{\sigma}] = \bar{\sigma}G_K,$$

where $\bar{\sigma}$ is any extension of σ to all of E. (Such an extension $\bar{\sigma}$ exists for every σ, since E is a normal extension of F. Notice that $[\bar{\sigma}] = \{\bar{\sigma} \in G(E/F) | \bar{\sigma} \restriction K = I\}$.) We have to prove the following facts about ψ:

(a) ψ is well defined: If $\bar{\sigma}_1$ and $\bar{\sigma}_2$ are two extensions of σ, then $[\bar{\sigma}_1] = [\bar{\sigma}_2]$, so that for each $\sigma \in G(K/F)$ there is a unique $\psi(\sigma)$.

(b) ψ is one to one.

(c) ψ maps $G(K/F)$ onto all of G_F/G_K.

(d) ψ is a homomorphism.

To show (a): Suppose $\bar{\sigma}_1$ and $\bar{\sigma}_2$ are both extensions of σ, so that for every $x \in K$ we have $\sigma(x) = \bar{\sigma}_1(x) = \bar{\sigma}_2(x)$. Let $\varphi = \bar{\sigma}_2^{-1}\bar{\sigma}_1$ and $x \in K$. Remembering that $\sigma(x) \in K$, we have

$$\varphi(x) = (\bar{\sigma}_2^{-1}\bar{\sigma}_1)(x) = \bar{\sigma}_2^{-1}(\bar{\sigma}_1(x)) = \bar{\sigma}_2^{-1}(\sigma(x)) = \sigma^{-1}(\sigma(x))$$

$$= (\sigma^{-1}\sigma)(x) = x,$$

which shows that φ leaves every element of K fixed, so $\varphi \in G_K$. Therefore, $\bar{\sigma}_1 = \bar{\sigma}_2\varphi \in \bar{\sigma}_2G_K$, and $\bar{\sigma}_1 \in [\bar{\sigma}_2]$, which implies that $[\bar{\sigma}_1] = [\bar{\sigma}_2]$, because cosets are either identical or disjoint.

To show (b): In part (a) we showed that for each σ there is only one $\psi(\sigma)$; now we must show that each $\psi(\sigma)$ comes from only one $\sigma \in G(K/F)$, in other words, that

$$\psi(\sigma_1) = \psi(\sigma_2) \Rightarrow \sigma_1 = \sigma_2, \qquad \forall \sigma_1, \sigma_2 \in G_K(K/F).$$

So suppose $\psi(\sigma_1) = \psi(\sigma_2)$ or, equivalently, $[\bar{\sigma}_1] = [\bar{\sigma}_2]$. Then $\bar{\sigma}_1 \in [\bar{\sigma}_2]$, so there is some $\varphi \in G_K$ such that $\bar{\sigma}_1 = \bar{\sigma}_2\varphi$. For every $x \in K$ we have, therefore,

$$\sigma_1(x) = \bar{\sigma}_1(x) = (\bar{\sigma}_2\varphi)(x) = \bar{\sigma}_2(\varphi(x)) = \bar{\sigma}_2(x) = \sigma_2(x),$$

since $\varphi(x) = x$ for every $x \in K$ and $\sigma_1 = \bar{\sigma}_1 \upharpoonright K$, $\sigma_2 = \bar{\sigma}_2 \upharpoonright K$, by the definition of $\bar{\sigma}_1, \bar{\sigma}_2$. The last equation shows that $\sigma_1 = \sigma_2$, as we wanted.

(c) To show that ψ is onto G_F/G_K: Suppose we are given some coset $\bar{\sigma}G_K \in G_F/G_K$, with given $\bar{\sigma} \in G_F$. We must show that there is some $\sigma \in G(K/F)$ such that $\psi(\sigma) = \bar{\sigma}G_K$. Since $G_K \lhd G_F$ we know that $K \rhd F$, so $\bar{\sigma}(K) = K$ for every $\bar{\sigma} \in G_F$, by the definition of normality. Therefore, $\bar{\sigma}$ is an extension of some $\sigma \in G(K/F)$, and for this σ we do indeed have $\psi(\sigma) = \bar{\sigma}G_K$.

(d) ψ is a homomorphism: We must show that

$$\psi(\sigma_1) \cdot \psi(\sigma_2) = \psi(\sigma_1 \cdot \sigma_2).$$

Let

$$\psi(\sigma_1) = [\bar{\sigma}_1],$$

$$\psi(\sigma_2) = [\bar{\sigma}_2],$$

$$\psi(\sigma_1 \cdot \sigma_2) = [\bar{\sigma}_1\bar{\sigma}_2].$$

Then $\psi(\sigma_1) \cdot \psi(\sigma_2) = [\bar{\sigma}_1][\bar{\sigma}_2] = [\bar{\sigma}_1\bar{\sigma}_2] = \psi(\sigma_1\sigma_2)$, since $G_K \lhd G_F$ so that cosets can be multiplied, which proves that ψ is a homomorphism and completes the proof of part (3).

Proof of Part (4). This is essentially just a restatement of the last theorem of Section 2.4, for this says that $[E:K] = |G_K|$, and from part (3) we see that

$$[K:F] = |G(K/F)| = \frac{|G_F|}{|G_K|} = \mathrm{ind}_{G_F} G_K. \quad \|$$

3.4 FUNDAMENTAL THEOREM

This completes the proof of the fundamental theorem of Galois theory. In Chapter IV we shall describe some of the many applications of the results we obtained and give one method for determining the group of a polynomial and one for carrying out the actual solution of those equations which are solvable in terms of radicals.

EXERCISE. Suppose $E \rhd Q$ with $G(E/Q)$ Abelian and that $F < E$. Show that $F \rhd Q$ and that $G(F/Q)$ is also Abelian.

CHAPTER IV

APPLICATIONS

4.1 SOLVABILITY OF EQUATIONS

The fundamental theorem establishes a correspondence between fields and their Galois groups. We now wish to apply this relation to the problem of solving algebraic equations. Intuitively speaking, an equation $f(x) = 0$ is solvable if we can find some expression α built up of elements of the coefficient field F by means of $+, \cdot$, and $\sqrt[k]{\ }$ in a finite number of steps such that $f(\alpha) = 0$. Two questions arise immediately:

(1) Shall we require a constructive procedure for obtaining α?

(2) Or is it enough to assert the existence of α?

To make matters more precise, we introduce the following definitions:

DEFINITION.   A polynomial $f(x) \in F[x]$ is called *solvable* over F if there is a finite sequence of fields $F = F_0 < F_1 < \cdots < F_k$ and a finite sequence of integers $n_0, \ldots, n_{k-1}$ such that $F_{i+1} = F_i(\gamma_i)$, with $\gamma_i^{n_i} \in F_i$, and if all the roots of $f(x)$ lie in F_k, that is, $E \subseteq F_k$, where E is the splitting field of $f(x)$.

Such a sequence of fields $F_0 < \cdots < F_k$ is called a *root tower* over F_0 or an *extension of F_0 by radicals*, or often, more briefly, *tower over F_0* or *radical extension of F_0*. Sometimes the field F_k itself is called a *radical extension of F_0*.

It is conceivable that there might be polynomials $f(x)$ with roots $\alpha_1, \ldots, \alpha_n$

such that α_1 is in some root tower over F but $\alpha_2, \ldots, \alpha_n$ are not, so one root of $f(x)$ may be expressible in radicals but not all the roots. This situation is however excluded by the following theorem:

THEOREM. If char $F = 0$ and $f(x)$ is irreducible over $F[x]$ and there is a root tower for one root α of $f(x)$, then there is a tower over F which contains all the roots of $f(x)$. (See Section 4.2, p. 134.)

DEFINITION. A polynomial $f(x) \in F[x]$ is called *explicitly solvable* over F if $f(x)$ is solvable over F and if, in addition, there is a constructive procedure which after finitely many steps enables us to write down $\alpha_1, \alpha_2, \ldots, \alpha_n \in F_k$ such that

$$f(x) = (x - \alpha_1)(x - \alpha_2) \cdots (x - \alpha_n).$$

The second definition is evidently stronger than the first. It insists not only on the existence of an algebraic extension F_k of F in which the roots are contained but that these roots should be found and expressed in terms of radicals. It will turn out, however, that the two definitions are generally equivalent. In fact, this is true whenever char $F > \deg f(x)$ and every polynomial over F can be factored into irreducible factors in a finite number of steps. If F is denumerable, this can easily be proved as follows. Suppose $f(x) = 0$ is solvable over F. If F is denumerable, so is the algebraic closure $\mathscr{F}$ of F. Let $\gamma_1, \gamma_2, \ldots$ be an enumeration of the elements of $\mathscr{F}$. Assuming that $f(x)$ is solvable, let γ_i be the first element in this enumeration for which $f(x)$ vanishes, so that $f(\gamma_i) = 0$. Let $\alpha_1 = \gamma_i$ and consider next the polynomial

$$f_1(x) = \frac{f(x)}{x - \gamma_i} = \frac{f(x)}{x - \alpha_1} \in \mathscr{F}[x].$$

We already know that $\gamma_1, \ldots, \gamma_{i-1}$ are not roots of f_1, for otherwise they would also be roots of $f(x)$, but α_1 might be a double root of $f(x)$ and therefore also a root of $f_1(x)$. Remember that it is not necessarily evident on inspection whether two given elements $\alpha, \beta \in \mathscr{F}$ are equal: For example, $\sqrt{2} + \sqrt{3} = \sqrt{5 + \sqrt{24}}$ and $\sqrt{2} - \sqrt{3} = $

$-\sqrt{5-\sqrt{24}}$. Since $f(x)$ is solvable, so is $f_1(x)$, assuring us that there will be a first γ_j in the sequence $\gamma_i, \gamma_{i+1}, \ldots$ for which $f_1(\gamma_j) = 0$. Let this γ_j be α_2 and continue with the polynomial

$$f_2(x) = \frac{f_1(x)}{x - \alpha_2} \in \mathscr{F}[x].$$

The degree of $f(x)$ is finite and the factorization of polynomials over a field is unique, so this procedure must terminate when we have found n numbers $\alpha_1, \ldots, \alpha_n \in \mathscr{F}$ such that $f_i(\alpha_{i+1}) = 0$. Referring to the definitions of $f_i(x)$, we get

$$f(x) = (x - \alpha_1) \cdots (x - \alpha_n).$$

This process is quite clearly not the one mathematicians had in mind when they were searching for a procedure to solve equations. It is aesthetically unattractive and totally impractical, but it does give an existence proof of sorts. By means of groups and Galois theory it is possible to devise a method that leads directly to the correct result whenever the given equation is solvable at all and also tells us which equations are solvable and which are not. More precisely, knowing the group G of an equation, we next find a composition series of G. The given equation will then be solvable if and only if this series has all its factors of prime order; moreover, knowing these groups will tell us exactly how to go about solving the equation.

It is therefore apparent that once we know the group of a polynomial we can examine it, find its composition series and composition factors, determine whether the group is solvable, and, if so, go about the solution of the equation. This leaves us the big question: Given $f(x) \in F[x]$, how do we find its group? Luckily there is also a constructive procedure for this. It is described in Section 4.9.

As everyone knows, all equations of degree ≤ 4 are explicitly solvable over Q, and by the same methods over any field F of characteristic 0. This is not the case when the characteristic is not zero.

Let $F = GF(2)$ and let us try to solve the equation

$$f(x) = x^2 + x + 1 = 0.$$

4.1 SOLVABILITY OF EQUATIONS

The formula $x = -\frac{1}{2} \pm \frac{1}{2}\sqrt{-3}$ cannot be used here, because 2 appears in the denominator and we are working mod 2. In addition, $f(x)$ has no roots in F and is therefore irreducible. [As a check we get $f(1) = $ _____, $f(0) = $ _____.] Let α be a root of $f(x) = 0$. Then $F(\alpha) \cong F[x]/(x^2 + x + 1) \cong GF(2^2)$, and $[F(\alpha)/F] = 2$, so $G(F(\alpha)/F)$ is the cyclic group of order 2. Therefore, if $f(x)$ were solvable by radicals, we would be able to find a field $F(\sqrt{m})$ with $m \in F$ such that $\alpha \in F(\sqrt{m})$ and we could write α in the form

$$\alpha = a + b\sqrt{m},$$

with a, b, and m in F. The only possibilities for m in this case are $m = 0$ or $m = 1$, neither of which is of any help. The equation is therefore unsolvable.

Incidentally, let us point out here that there is no primitive square root of unity mod 2. For the equation $x^2 = 1$ has 1 as a double root, and 1 is not a primitive square root of unity. Similarly, there is no primitive cube root of unity mod 3: the equation $x^3 = 1$ has 1 as a triple root, mod 3. (How about the general case: Can we ever have primitive pth roots of unity mod p?)

Suppose we try to remedy this defect by adjoining some new element β to $GF(2)$ with the properties one would like a primitive square root of unity to have: $\beta^2 = 1$ and $\beta \neq 1$. Is this allowable, or would this lead to trouble? How about adjoining a primitive pth root of unity to $GF(p)$?

It may also be worth pointing out that the assumption that every polynomial over F can be factored into irreducible factors in a finite number of steps is vital and cannot be reduced. To see this we simply consider the case when $F = R$, the real numbers. Every polynomial $f(x) \in R[x]$ is the product of linear and quadratic polynomials, and so is clearly solvable. The roots will always lie in $C = R(i)$, but there is no method for writing them down explicitly using only the coefficients of $f(x)$ and $+$, $-$, $\times$, $\div$, and $\sqrt[k]{\ }$ as operations. As an example, consider the polynomial $f(x) = x^5 - x^4 - x^2 + 2x - 1$, which is irreducible over Q but factors into three linear factors and one quadratic factor over R. (The last statement is true because

_____.)

It is possible to show that $f(x)$ is not solvable in radicals over Q. If we could factor in a constructive manner over R we would not need the many approximation methods used to find the roots of polynomials. We therefore come to the conclusion that every polynomial is solvable over R, but some are not explicitly solvable. This situation would have been avoided if we had adopted the perhaps still more intuitive notion for the solvability of $f(x)$ which would require that the field F_0 used as base of the root tower in the first definition of this section be the smallest field containing all the coefficients of $f(x)$ and had defined the group of $f(x)$ as the group of the splitting field E over this coefficient field F_0.

4.2 SOLVABLE EQUATIONS HAVE SOLVABLE GROUPS

We now start our work toward Galois's theorem that a polynomial is solvable if and only if its group is solvable. As we saw in Section 4.1, this result may fail if the characteristic of the ground field F is different from 0 and that "solvable" cannot be replaced by "explicitly solvable." In other words, for some fields of characteristic 0 such as R, Galois's theorem is only an existence theorem: We can prove the existence of an extension by radicals which must contain the roots of $f(x)$, but there is no constructive method for exhibiting these roots. For the real field R we can, of course, exhibit the splitting field very easily. By the fundamental theorem of algebra it must be either R(i) or R itself, and by using one of the methods for determining the number of real roots of $f(x)$ (for example, by graphing or using Sturm's theorem [W, p. 82]) we can decide which of the two it is.

From now on we shall assume that the characteristic of the fields involved is zero and shall only occasionally indicate briefly how the argument and conclusions would have to be changed in case the characteristic is finite. Most of the theorems hold whenever the characteristic is "sufficiently large," and a careful examination of the proof will show that $n!$ is generally "sufficiently large" if the polynomial in question is of degree n.

4.2   SOLVABLE EQUATIONS HAVE SOLVABLE GROUPS

The definition in Section 4.1 states that a polynomial is solvable if and only if its splitting field E is contained in some extension F_k of F_0 by radicals. The sequence of fields $F_0, \ldots, F_k$ is not uniquely determined by $f(x)$, as we see in the following examples of a few specific radical extensions:

EXAMPLES.   (1) The polynomial

$$f(x) = x^4 - 2x^2 - 8x + 1$$

has one root $\alpha = \sqrt{2} + \sqrt{3}$, its splitting field $E = Q(\sqrt{2}, \sqrt{3})$, and $F_0 = Q$.

We can form the following radical extensions (many others are also possible):

(a) $Q < Q(\sqrt{2}) < Q(\sqrt{2}, \sqrt{3}) = E,$

(b) $Q < Q(\sqrt{3}) < Q(\sqrt{2}, \sqrt{3}) = E,$

(c) $Q < Q(i) < Q(\sqrt{i}) < Q(\sqrt{i}, \sqrt{3}) \gneqq E,$

(d) $Q < Q(\sqrt{3}) < Q(\sqrt{3}, \sqrt{5}) < Q(\sqrt{3}, \sqrt{5}, \sqrt{2}) \gneqq E,$

(e) $Q < Q(\sqrt{6}) < Q(\sqrt{6}, \sqrt{2}) = E,$

(f) $Q < Q(\sqrt{6}) < Q(\sqrt{6}, \sqrt{3}) = E,$

(g) $Q < Q(\sqrt{6}) < Q(\sqrt{6}, \sqrt{5 + 2\sqrt{6}}) = E.$

[One should check that $\sqrt{2} \in Q(\sqrt{i})$, because _____

_____, and that $Q(\sqrt{6}, \sqrt{5 + 2\sqrt{6}}) = Q(\sqrt{5 + 2\sqrt{6}}) =$

$Q(\sqrt{2} + \sqrt{3}) = Q(\sqrt{2}, \sqrt{3})$, because _____

_____.]

(2) The polynomial

$$f(x) = x^3 + 3x - 2$$

has roots

$$\alpha_1 = \sqrt[3]{1 + \sqrt{2}} + \sqrt[3]{1 - \sqrt{2}},$$

$$\alpha_2 = \text{\underline{\hspace{4cm}}},$$

$$\alpha_3 = \text{\underline{\hspace{4cm}}},$$

with $E = Q(\alpha_1, \alpha_2, \alpha_3)$ and $F_0 = Q$. We can form the following radical extensions:

(a) $Q < Q(\sqrt{2}) < Q(\sqrt{2}, \sqrt[3]{1 + \sqrt{2}}) < Q(\sqrt{2}, \sqrt[3]{1 + \sqrt{2}}, \sqrt[3]{1 - \sqrt{2}})$
$< Q(\sqrt{2}, \sqrt[3]{1 + \sqrt{2}}, \sqrt[3]{1 - \sqrt{2}}, \omega) \geq E$,

(b) $Q < Q(\omega\sqrt{2}) < Q(\omega\sqrt{2}, \sqrt[3]{1 + \sqrt{2}}) \geq E$.

[The field $Q(\omega\sqrt{2}, \sqrt[3]{1 + \sqrt{2}})$ is of the form $Q(\omega\sqrt{2}, \beta)$ with $\beta^3 \in Q(\omega\sqrt{2})$, because $\sqrt{2} = \frac{1}{2}(\omega\sqrt{2})^3 \in Q(\omega\sqrt{2})$.]

(3) The polynomial

$$f(x) = x^3 - 3x + 1$$

has roots

$$\alpha_1 = (\omega^{1/3} + \omega^{-1/3}),$$

$$\alpha_2 = (\omega^{2/3} + \omega^{-2/3}),$$

$$\alpha_3 = (\omega^{4/3} + \omega^{-4/3}),$$

with $F_0 = Q$.

We can form the following radical extension:

$$Q < Q(\omega) < Q(\omega^{1/3}) \gneqq E.$$

In this case $E = Q(\alpha_1, \alpha_2, \alpha_3)$ is certainly not the same as $Q(\omega^{1/3})$. This is not obvious, but one can show that $G(E/Q) = \mathfrak{A}_3$, so $|G(E/Q)| = 3$, while $[Q(\omega^{1/3}):Q] = 6 \neq 3$.

Before we can prove that solvable equations have solvable groups, we shall need a few preliminary theorems.

THEOREM. If ε is a primitive nth root of unity and B is a field, then $B(\varepsilon)$ is an Abelian normal radical extension of B and $G(Q(\varepsilon)/Q)$ is isomorphic to the multiplicative group of integers mod n, $Z_n^\times$. (Remember $k \in Z_n^\times$ only if $\gcd(k, n) = 1$.)

4.2 SOLVABLE EQUATIONS HAVE SOLVABLE GROUPS

Proof. (1) Since $\varepsilon^n = 1$ and $1 \in B$, we see that $B(\varepsilon)$ is a radical extension of B. [It does not really seem fair to call $B(\sqrt[n]{1})$ a radical extension of B, so in Section 4.4 we shall show how one can find all n different values of $\sqrt[n]{1}$ and express them in terms of radicals of lower degree. DeMoivre's theorem gives the values immediately in terms of trigonometric functions.]

(2) Let B^* be the splitting field of $(x^n - 1)$ over B, and let $\varepsilon_1, \ldots, \varepsilon_n$ be all the nth roots of unity. Then ε is one of the $\varepsilon_1, \ldots, \varepsilon_n$, so $B(\varepsilon) < B^*$. However, ε is assumed primitive, by hypothesis, so each ε_i is of the form ε^k, for some integer k. Therefore, $B^* = B(\varepsilon_1, \ldots, \varepsilon_n) < B(\varepsilon)$, showing that $B^* = B(\varepsilon)$. Moreover, B^* is the splitting field of $x^n - 1$, so $B \lhd B^*$, that is, $B \lhd B(\varepsilon)$.

(3) Let $\sigma, \tau \in G(B(\varepsilon)/B)$. We must show that $\sigma(\tau(\alpha)) = \tau(\sigma(\alpha))$ for every $\alpha \in B(\varepsilon)$. It is sufficient to show that $\sigma(\tau(\varepsilon)) = \tau(\sigma(\varepsilon))$, because _____

_____. We know that ε is an nth root of unity, so $\tau(\varepsilon)$ must also be an nth root, since _____

_____.

(4) Therefore, $\tau(\varepsilon) = \varepsilon^k$, for some k. Similarly, $\sigma(\varepsilon) = \varepsilon^l$, for some l. Combining these results we get $\sigma(\tau(\varepsilon)) = \sigma(\varepsilon^k) = (\sigma(\varepsilon))^k = (\varepsilon^l)^k = \varepsilon^{lk} = (\varepsilon^k)^l = (\tau(\varepsilon))^l = \tau(\varepsilon^l) = \tau(\sigma(\varepsilon))$, which proves that the group is Abelian.

(5) That $G(Q(\varepsilon)/Q) \cong Z_n$ follows from the exercises below. $\|$

EXERCISES. Prove that if ε is a primitive nth root of unity and $\varepsilon \notin B$, then

(1) ε^k is a primitive nth root of unity if and only if $\gcd(k, n) = 1$.

(2) If $\sigma \in G(B(\varepsilon)/B)$ and ε is primitive, then $\sigma(\varepsilon)$ is primitive.

(3) $G(B(\varepsilon)/B)$ is isomorphic to a subgroup of $Z_n^{\times}$, where $Z_n^{\times}$ is the group whose elements are the integers j for which $1 \le j \le n - 1$ and $\gcd(j, n) = 1$, with multiplication mod n as the group operation [Ad, p. 131].

We shall return to the nth roots of unity in Section 4.4.

THEOREM. If char $B = 0$, b is an element of the field B and B^* is the splitting field of $(x^n - b)$ over B, then $B \lhd B^*$ and $G(B^*/B)$ is solvable.

Proof. (1) Clearly $B \lhd B^*$, for B^* is by definition the splitting field of $x^n - b$ over B.

(2) The group $G(B^*/B)$ is therefore the Galois group of $x^n - b$. It is not generally Abelian, for we saw that the group of $x^3 - 2$ is $\mathfrak{S}_3$, which is not Abelian.

(3) To show that $G(B^*/B)$ is nonetheless solvable, we must therefore find a composition series with Abelian factors. We shall show that the sequence of fields

$$B < B(\varepsilon) < B^*$$

is a suitable series, whenever $b \neq 0$ and where ε is a primitive nth root of unity. (If $b = 0$, then $B^* = B$ and the theorem holds trivially.)

(4) Since $B^* \rhd B$, we also have $B^* \rhd B(\varepsilon)$. Also $B(\varepsilon) \rhd B$, because it is the splitting field of $x^n - 1$.

(5) We therefore have a sequence of normal extensions and we list the corresponding group below each field:

$$\left\{ \begin{array}{l} B \qquad \lhd B(\varepsilon) \qquad \lhd B^*, \\[2mm] G(B^*/B) \rhd G(B^*/B(\varepsilon)) \rhd G(B^*/B^*) = (e). \end{array} \right.$$

(6) This sequence of groups forms a composition series, as each is a normal (but not necessarily maximal normal) subgroup of the preceding group. The factor groups are

$$G(B^*/B)/G(B^*/B(\varepsilon)) \quad \text{and} \quad G(B^*/B(\varepsilon))/(e) = G(B^*/B(\varepsilon)).$$

(7) By the fundamental theorem,

$$G(B^*/B)/G(B^*/B(\varepsilon)) \cong G(B(\varepsilon)/B)$$

and this is Abelian by the preceding theorem.

(8) To see that the second group is also Abelian, let $\beta, \varepsilon\beta, \ldots, \varepsilon^{n-1}\beta$ be the roots of $x^n - b$ and suppose that $\sigma, \tau \in G(B^*/B(\varepsilon))$. The automorphisms σ and τ therefore leave every element of $B(\varepsilon)$ fixed, so, in particular, $\sigma(\varepsilon) = \tau(\varepsilon) = \varepsilon$.

(9) Moreover, σ and τ can send a root of $x^n - b$ only into some other root of

4.2 SOLVABLE EQUATIONS HAVE SOLVABLE GROUPS

$x^n - b$, so there exist integers i and j $(0 \leq i, j \leq n - 1)$ such that $\sigma(\beta) = \varepsilon^i \beta$ and $\tau(\beta) = \varepsilon^j \beta$.

(10) Then for any root $\varepsilon^k \beta$ of $x^n - b$ we have $\sigma\tau(\varepsilon^k \beta) = \sigma(\tau(\varepsilon^k) \cdot \tau(\beta)) = \sigma(\varepsilon^k \cdot \varepsilon^j \beta) = \sigma(\varepsilon^{k+j}\beta) = \sigma(\varepsilon^{k+j}) \cdot \sigma(\beta) = \varepsilon^{k+j} \cdot \varepsilon^i \beta = \varepsilon^{k+j+i}\beta = \varepsilon^{k+i} \cdot \varepsilon^j \beta = \tau(\varepsilon^{k+i}) \cdot \tau(\beta) = \tau(\varepsilon^{k+i}\beta) = \tau(\varepsilon^k \cdot \varepsilon^i \beta) = \tau(\sigma(\varepsilon^k) \cdot \sigma(\beta)) = \tau\sigma(\varepsilon^k \beta)$.

(11) Therefore, σ and τ commute over all the roots of $x^n - b$ and therefore over all the elements of B^*, so $G(B^*/B(\varepsilon))$ is Abelian.

(12) So $G(B^*/B)$ has Abelian factor groups and is solvable.   ‖

EXERCISE.   Show that $G(B^*/B(\varepsilon)) \cong C_n$, the cyclic group with n elements.

Our object is to prove that the Galois group of every solvable polynomial is solvable. The hypothesis thus is that the splitting field E of $f(x)$ is contained in a radical extension K of the ground field F:

$$F \lhd E < K.$$

We can form $G(K/F)$ and show that this is solvable, but this does not tell us very much about $G(E/F)$, for in general unfortunately K is not a normal extension of E, nor of F. If it were, we could use the fundamental theorem of Galois theory and we would have the following situation:

$$F \lhd E \quad \lhd K \quad \text{and} \quad F \lhd K,$$

$$G(K/F) \rhd G(K/E) \rhd G(K/K) = (e),$$

and

$$G(K/F)/G(K/E) \cong G(E/F).$$

The group $G(E/F)$, the one we are after, would therefore be a homomorphic image of the solvable group $G(K/F)$ by the canonical mapping for quotient groups and so $G(E/F)$ would be solvable, which is what we want to prove. But if the radical extension

K is not a normal extension of F, this scheme will not work, for then we cannot use the fundamental theorem. To get around this, we shall use the next theorem to construct a new field containing E which is both a normal and a radical extension of F.

THEOREM. If $F \lhd F' < F'(\gamma)$, where $\gamma'' \in F'$, then we can find a field F'' such that
   (a) $F \lhd F' < F'(\gamma) \lhd F''$.
   (b) $F \lhd F''$.
   (c) F'' is a radical extension of F'.
   (d) $G(F''/F')$ is solvable.
   (e) If $G(F'/F)$ is solvable; then $G(F''/F)$ is solvable.

Before we give the proof we shall go through an example to illustrate the method.

EXAMPLE.   Let

$$F = Q,$$

$$F' = Q(\omega, \sqrt[3]{2}),$$

$$\gamma = \sqrt[4]{1 + \sqrt[3]{2}}.$$

Then $n = 4$, and $\gamma'' = 1 + \sqrt[3]{2} \in F'$. We have $F'(\gamma) = Q(\omega, \sqrt[3]{2}, \gamma) = Q(\omega, \gamma)$, because $\sqrt[3]{2} = \gamma^4 - 1$. If we let ε be a primitive nth root of unity (in our example here, therefore, $\varepsilon = +i$ or $-i$), then

$$F \lhd F' \lhd F'(\gamma, \varepsilon),$$

but one can show that $F'(\gamma, \varepsilon) = Q(\omega, i, \sqrt[4]{1 + \sqrt[3]{2}})$ is not a normal extension of F, so part (b) of the statement of the theorem would not be satisfied by taking $F'' = F'(\gamma, \varepsilon)$. To correct this situation we first enlarge $F'(\gamma)$ as follows.

We know that $\gamma^4 \in F'$. Let us look at all the conjugates of γ^4 in F' under the

4.2 SOLVABLE EQUATIONS HAVE SOLVABLE GROUPS

automorphisms of $G(F'/F)$. They are (including γ^4 itself)

$$\alpha_1 = \gamma^4 = 1 + \sqrt[3]{2},$$

$$\alpha_2 = \eta^4 = 1 + \omega\sqrt[3]{2},$$

$$\alpha_3 = \zeta^4 = 1 + \omega^2\sqrt[3]{2},$$

all of them roots of the polynomial $p(x) = $ _____ $\in Q[x]$.
Therefore, γ is a root of the polynomial $q(x) = p(x^4) = $ _____
$\in Q[x]$, and if we let F'' be the splitting field of $q(x)$, then $F'' = F'(\varepsilon, \gamma, \eta, \zeta_i)$ and

(1) $Q \lhd F' < F'(\gamma) \lhd F''$,

(2) $Q \lhd F''$,

(3) F'' is a radical extension of F'.

Parts (a) and (b) follow because F'' is the splitting field of $g(x)$ over Q and part (c)
follows because $F'' \rhd F'(\varepsilon, \gamma, \eta) \rhd F'(\varepsilon, \gamma) \rhd F'(\varepsilon) \rhd F'$ is an extension by radicals
of F'. To see that parts (d) and (e) hold we have the two corresponding sequences

$$F \lhd \quad F' \quad \lhd \quad F'(\varepsilon) \quad \lhd \quad F'(\varepsilon, \gamma) \quad \lhd F'(\varepsilon, \gamma, \eta)$$

$$\lhd F'(\varepsilon, \gamma, \eta, \zeta) = F'',$$

$$G(F''/F) \rhd G(F''/F') \rhd G(F''/F'(\varepsilon)) \rhd G(F''/F'\varepsilon, \gamma)) \rhd G(F''/F'(\varepsilon, \gamma, \eta))$$

$$\rhd G(F''/F'') = (e).$$

Since we now have $F \lhd F''$, we may use the fundamental theorem and we see that the
composition factors are isomorphic to

$$G(F'/F), \quad G(F'(\varepsilon)/F'), \quad G(F'/\varepsilon, \gamma)/F'/(\varepsilon)), \quad G(F'(\varepsilon, \gamma, \eta)/F'(\varepsilon, \eta)), \quad G(F''/F'(\varepsilon, \gamma, \eta)),$$

respectively. From Chapter III we know that $G(F'/F) \cong \mathfrak{S}_3$, which is solvable, and
the preceding theorems show that each of the other factors is Abelian, so $G(F''/F)$ is
solvable. [If we had inserted $Q(\omega)$ as an additional intermediate field between F and
F' all factors would have been Abelian. Also note that in this example $F'(\varepsilon, \gamma, \eta) =$
$F'(\varepsilon, \gamma, \eta, \zeta) = F''$.]

Proof of the Theorem. The proof of the general theorem is merely a generalization of the procedure used in this example:

(1) By hypothesis there is an element α_1 in F' such that $\gamma^n = \alpha_1$.

(2) Let $\alpha_1, \alpha_2, \ldots, \alpha_k$ be all the conjugates of α_1 in F'; that is, they are all possible elements of the form $\sigma\alpha_1$ for $\sigma \in G(F'/F)$. Also let ε be a primitive nth root of unity and let γ_i be such that $\gamma_i^n = \alpha_i$, setting $\gamma_1 = \gamma$. Let

$$F'' = F'(\varepsilon, \gamma_1, \ldots, \gamma_k).$$

(3) We then have a tower

$$F' \lhd F'(\varepsilon) \lhd F'(\varepsilon, \gamma_1) \lhd F'(\varepsilon, \gamma_1, \gamma_2) \lhd \cdots \lhd F'(\varepsilon, \gamma_1, \gamma_2, \ldots, \gamma_k) = F''.$$

(4) Each of these extensions is a normal extension of the preceding, because it is obtained by adjoining all the roots of a separable polynomial: $F'(\varepsilon)$ from $(x^n - 1)$, $F'(\varepsilon, \gamma_1)$ from $(x^n - \alpha_1)$, and so on.

(5) Therefore, (a) and (c) hold, and if we let $F_i = F'(\varepsilon, \gamma_1, \ldots, \gamma_i)$, then $F_i \lhd F_{i+1}$ for $i = 0, \ldots, k - 1$.

(6) To show that part (b) holds, we show that F'' is the splitting field of the polynomial

$$p(x) = \prod_{i=1}^{k} (x^n - \alpha_i)$$

over F and that $p(x) \in F[x]$.

(7) Clearly $p(x)$ has all of its kn roots in F'', because each root of $p(x)$ is of the form $\varepsilon^h \gamma_i$ for some h and i.

(8) Also F'' is by definition generated by these roots, so it must be the splitting field of $p(x)$.

(9) Each coefficient of $p(x)$ is a symmetric function of $\alpha_1, \ldots, \alpha_k$. Since every automorphism $\alpha \in G(F'/F)$ permutes $\alpha_1, \ldots, \alpha_k$ among themselves, the coefficients of $p(x)$ are all invariant under the automorphisms of $G(F'/F)$.

(10) They are therefore all elements of F and so $p(x) \in F[x]$.

4.2 SOLVABLE EQUATIONS HAVE SOLVABLE GROUPS

(11) Combining steps (8) and (10) shows that $F \lhd F''$.

(12) We now go back to the sequence of fields

$$F \quad \lhd \quad F' \quad \lhd \quad F'(\varepsilon) \quad \lhd \quad F_1 \quad \lhd F_2 \lhd \cdots \lhd \quad F_k = F''$$

and the corresponding sequence of groups

$$G(F''/F) \rhd G(F''/F') \rhd G(F''/F'(\varepsilon)) \rhd G(F''/F_1) \rhd \quad \cdots \quad \rhd G(F''/F_k) = (e).$$

(13) Since $F \lhd F''$, we can apply the fundamental theorem and get the quotient groups

$$G(F''/F)/G(F''/F'), \quad G(F''/F')/G(F''/F'(\varepsilon)), \quad G(F''/F'(\varepsilon))/F(F''/F_1), \ldots,$$

$$G(F''/F_{k-1})/G(F''/F_k),$$

which are isomorphic, respectively, to

$$G(F'/F), \quad G(F'(\varepsilon)/F'), \quad G(F_1/F'(\varepsilon)), \ldots, G(F_{k-1}/F_k).$$

(14) From the second one of these on they are all Abelian, by the preceding theorems, and form a sequence of composition factors for $G(F''/F')$.

(15) Therefore, $G(F''/F')$ is solvable, proving (d).

(16) If, in addition, $G(F'/F)$ is solvable, then also $G(F''/F)$ is solvable so (e) holds.

(17) This completes the proof.   ‖

The next theorem may somewhat loosely be stated as "every radical extension of a solvable extension has a solvable group":

THEOREM. If $F \lhd F_1 < B$, where F_1 is a solvable extension of F and B is a radical extension of F_1, then there is a normal radical extension K of F such that $B \lhd K$ and $G(K/F)$ is solvable. (Here F_1 may, but need not, equal F.)

Proof. To prove this we shall simply make repeated applications of the preceding theorem:

(1) We are given that there is a finite sequence of fields

$$F \lhd F_1 < F_2 < F_3 < \cdots < F_k = B$$

such that $F_{i+1} = F_i(\gamma_i)$ and $\gamma_i^{n_i} \in F_i$ for some integer n_i.

(2) By the last theorem, we can find a normal radical extension F_1'' of F such that

$$F \lhd F_1 \lhd F_1'' \qquad \text{and} \qquad F \lhd F_1''.$$

(3) Let $F_2' = F_1''(\gamma_1)$, so F_2' is a radical extension of F_1''.

(4) Again by the last theorem, there is a normal radical extension F_2'' of F_2' whose group over F is solvable.

(5) We continue this process until we reach the field B and form the corresponding normal radical extension K of F with solvable group $G(K/F)$. ‖

COROLLARY. If $F < B$ and B is a radical extension of F, then there is a normal radical extension K of F such that $B < K$ and $G(K/F)$ is solvable.

Proof. Take $F = F_1$ in the preceding theorem.   ‖

COROLLARY. If one root of $f(x) \in F[x]$ is expressible in radicals, then so are all the roots of $f(x)$.

Proof. (1) Let $\alpha_1, \ldots, \alpha_n$ be the roots of $f(x)$ and let B be a radical extension of F such that $F < F(\alpha_1) < B$.

(2) Then there is a normal radical extension K of F such that $F < F(\alpha_1) < B < K$.

(3) Since $K \rhd F$, all the roots $\alpha_1, \ldots, \alpha_n$ must be in K.   ‖

So far we have shown that if $f(x)$ is a polynomial over F with all roots expressible in terms of radicals over F, then its splitting field E can be embedded in a normal radical extension K of F and that $G(K/F)$ is solvable. We must still show that $G(E/F)$ is also solvable. The argument is very simple, and is essentially the same that we saw at the beginning of this section:

4.3   GENERAL EQUATION OF DEGREE n

THEOREM.   If $f(x) \in F[x]$ is solvable and E is its splitting field, then $G(E/F)$ is solvable.

Proof.   (1) By the last theorem, there is a normal radical extension K of F such that

$$F \lhd E \lhd K \qquad \text{and} \qquad F \lhd K$$

and $G(K/F)$ is solvable.

(2) Applying the fundamental theorem again, we get

$$G(K/F)/G(K/E) \cong G(E/F).$$

(3) This shows that $G(E/F)$ is a homomorphic image of $G(K/F)$ with kernel $G(K/E)$.

(4) Since the homomorphic image of a solvable group is solvable, this shows that $G(E/F)$ is solvable. $\|$

COROLLARY.   Let $f(x) \in F[x]$ with E as its splitting field. If $G(E/F)$ is not solvable, then $f(x)$ is not solvable in radicals over F. $\|$

4.3   GENERAL EQUATION OF DEGREE n

If there were a formula that would list the roots of any polynomial of degree n in terms of the coefficients, then this would also describe the roots of the equation

$$x^n + c_{n-1}x^{n-1} + \cdots + c_1 x + c_0 = 0,$$

when there is no algebraic relation whatever among the coefficients $c_{n-1}, \ldots, c_0$ and these are treated as formal entities. The ground field F in this case is $Q(c_0, \ldots, c_{n-1})$, where $c_0, \ldots, c_{n-1}$ are *indeterminates* (sometimes called *transcendentals*). This equation is then called the *general equation* of degree n. We shall show that its group is $\mathfrak{S}_n$.

THEOREM.   If $c_0, \ldots, c_{n-1}$ are transcendental over F, then the group of the equation

$$x^n + c_{n-1}x^{n-1} + \cdots + c_0 = 0$$

is the symmetric group $\mathfrak{S}_n$.

Proof. We shall prove this in three steps:

(1) If there were some algebraic relation among the roots $\alpha_1, \ldots, \alpha_n$ [in other words, suppose there were some polynomial $g(x_1, \ldots, x_n) \in Q[x_1, \ldots, x_n]$ such that $g(\alpha_1, \ldots, \alpha_n) = 0$], then there would be an algebraic relation among the coefficients $c_0, \ldots, c_{n-1}$.

(2) If there is no relation among the coefficients, then there is none among the roots.

(3) If there is no relation among the roots, then the group is $\mathfrak{S}_n$.

To prove (1): Suppose we had some nontrivial relation

$$g(\alpha_1, \alpha_2, \ldots, \alpha_n) = 0$$

among the roots $\alpha_1, \ldots, \alpha_n$. Let $\sigma \in \mathfrak{S}_n$ and let $\sigma g = g(\alpha_{\sigma(1)}, \alpha_{\sigma(2)}, \ldots, \alpha_{\sigma(n)})$, so that

$$(1)g = g(\alpha_1, \ldots, \alpha_n) = 0.$$

Then

$$\prod_{\sigma \in \mathfrak{S}_n} \sigma g = 0,$$

because at least one of its factors (namely $(1)g$) vanishes.

However, if τ is a fixed element of $\mathfrak{S}_n$, then

$$\{\sigma | \sigma \in \mathfrak{S}_n\} = \{\tau\sigma | \sigma \in \mathfrak{S}_n\},$$

and so

$$\prod_{\sigma \in \mathfrak{S}_n} \sigma g = \prod_{\sigma \in \mathfrak{S}_n} (\tau\sigma)g,$$

and since

$$\tau \prod_{\sigma \in \mathfrak{S}_n} \sigma g = \prod_{\sigma \in \mathfrak{S}_n} (\tau\sigma)g,$$

we see that $\prod_{\sigma \in \mathfrak{S}_n} \sigma g$ is a symmetric polynomial in $\alpha_1, \ldots, \alpha_n$. It is therefore expressible

4.3 GENERAL EQUATION OF DEGREE n

as a polynomial in the elementary symmetric functions of $\alpha_1, \ldots, \alpha_n$, that is, in the coefficients $c_0, \ldots, c_{n-1}$. This results in some polynomial $h(c_0, \ldots, c_{n-1})$ and we would have

$$\prod_{\sigma \in \mathfrak{S}_n} \sigma g = h(c_0, \ldots, c_{n-1}) = 0,$$

so $c_0, \ldots, c_{n-1}$ would not be algebraically independent.

Statement (2) is merely the contrapositive of (1) and follows immediately.

To prove (3): The splitting field of the equation is

$$E = F(c_0, \ldots, c_{n-1}, \alpha_1, \ldots, \alpha_n) = F(\alpha_1, \ldots, \alpha_n),$$

because $c_0, \ldots, c_{n-1}$ are symmetric polynomials in $\alpha_1, \ldots, \alpha_n$. Since by (2) there is no algebraic relation among $\alpha_1, \ldots, \alpha_n$, the mapping which sends α_i into $\alpha_{\sigma(i)}$ is an automorphism of $F(\alpha_1, \ldots, \alpha_n)$. Moreover, it leaves fixed every element of $F(c_0, \ldots, c_{n-1})$, so it is an element of the Galois group. As there is a one-to-one correspondence between such mappings and the elements σ of $\mathfrak{S}_n$, we see that the group of the equation must be $\mathfrak{S}_n$.   ‖

COROLLARY.   The general equation of degree n is not solvable for $n \geq 5$.

Proof.   The group is $\mathfrak{S}_n$ and this is not solvable (Section 1.2). By the results of the last section, this implies that the equation is not solvable.   ‖

This shows that there cannot be any general procedure which would lead to some formula for the roots of equations of degree 5 or higher, but it is not sufficient to show that there are equations with integer coefficients which are not solvable. The following remarkable theorem shows how we can construct such equations very easily. (For a proof of this theorem, see, for example, [A, p. 79; vdW, p. 189; or D, p. 295].)

THEOREM.   A solvable irreducible equation of prime degree over a field $F \subseteq \mathbf{R}$ has either exactly one real root or else all its roots are real.

Therefore, if we construct a polynomial $f(x)$ of degree 5 which has exactly three real roots, then we know immediately that the equation $f(x) = 0$ is not solvable. (Can one hope to find an irreducible polynomial of degree 5 with exactly two real roots? _____, because _____.)

One unsolvable polynomial is

$$x^5 - 4x + 2.$$

This polynomial is irreducible, because _____

_____, and it has exactly three real roots, because _____

_____.

Now that we have shown that there are unsolvable equations and, in fact, all equations with unsolvable groups are unsolvable, we are faced with two questions: (1) Is it true that all equations with solvable groups are solvable in radicals? (2) If we are given a specific equation, how can we determine whether its group is solvable or not? The next few sections will answer these questions.

4.4 ROOTS OF UNITY AND CYCLIC EQUATIONS

After showing that there do exist unsolvable equations, it is natural to look at the other side of the problem: Are all equations with solvable groups themselves solvable? If so, how does one go about finding the solution? As usual, before giving an answer to the general case, we should check first whether we can solve simple special cases. The special cases considered in this section are the equations whose group is cyclic and in particular those whose roots are the nth roots of unity.

Why should anyone go to all this trouble to evaluate the roots of unity in terms of radicals? Why not just simply write them as $\varepsilon = \sqrt[n]{1}$ and let it go at that? Suppose we did, how then could we write down the n different values of ε and tell them apart? We do want to be able to tell them apart, for once they are clearly distinguished from

4.4 ROOTS OF UNITY AND CYCLIC EQUATIONS

each other then we can also distinguish all the nth roots of any other element $a \in F$, simply by considering $\sqrt[n]{a}$ as denoting one specific nth root of a and writing the others as $\varepsilon^k \sqrt[n]{a}$, where $k = 1, 2, \ldots, n - 1$. We can take $\varepsilon = \cos(2\pi/n) + i\sin(2\pi/n)$, but this would involve trigonometric functions and these are surely not to be considered a solution in radicals, are unaesthetic in this context, and are inconvenient. The only solution is to hope that it is possible to express each nth root of unity ε in terms of radicals $\sqrt[r]{\ }$, where $r \lneqq n$. This is possible and we now proceed to do it. Again we shall restrict the discussion to fields of characteristic 0.

DEFINITION. The polynomial $f(x)$ and the equation $f(x) = 0$ are called *cyclic* if the Galois group of $f(x)$ is cyclic.

DEFINITION. The complex number ε is a *primitive* nth root of unity if $\varepsilon^n = 1$ and $\varepsilon^k \neq 1$ for $1 \leq k < n$. The monic polynomial of lowest possible degree $\Phi_n(x)$ among whose roots are all the primitive nth roots of unity is called the nth *cyclotomic polynomial*.

We have the following basic results:

THEOREM. (1) The coefficients of the cyclotomic polynomial are integers; that is,

$$\Phi_n(x) \in Z[x].$$

(2) The polynomial $\Phi_n(x)$ is irreducible over Q.

(3) If $\varphi(n)$ is the Euler φ function (the number of positive integers less than n and prime to n), then

$$\deg \Phi_n(x) = \varphi(n).$$

Proof. (1) Every nth root of unity is a root of $(x^n - 1)$, so suppose ε is a primitive nth root and is a root of the irreducible polynomial $h_\varepsilon(x)$. Then $h_\varepsilon(x)|(x^n - 1)$, because _____. Moreover, every other root ε' of $h_\varepsilon(x)$ is also a primitive nth root, because _____

_____. Therefore, $h_\varepsilon(x)|\Phi_n(x)$. Since $h_\varepsilon(x)|(x^n - 1)$, its leading coefficient is 1 and the other coefficients are all integers, by Gauss's theorem [Sec. 1.5 (16); a proof is given in MacD, p. 105]. If we let

$$k(x) = \text{lcm}\{h_\varepsilon(x)|\varepsilon \text{ is a primitive } n\text{th root}\},$$

then

$$k(x) = \Phi_n(x),$$

because _____

and

$$k(x)|(x^n - 1),$$

because _____.

Therefore, $k(x)$ has integer coefficients and leading coefficient 1.

(2) We make use of the theorem that a separable polynomial with no repeated factors is irreducible over F if and only if its Galois group over F expressed as a permutation group is transitive. (This theorem is proved below.)

If $h(x)$ is any irreducible factor of $\Phi_n(x)$, then $h(x)$ has root ε which is a primitive nth root, for, if not, then $\Phi_n(x)/h(x)$ would be a polynomial of lower degree than $\Phi_n(x)$ which also contained all primitive nth roots among its solutions.

Every root of $\Phi_n(x)$ is a primitive nth root: For if ε and η are roots of the irreducible factor $h(x)$, then $Q(\varepsilon) \cong Q(\eta)$, and $\eta^i = 1$ iff $\varepsilon^i = 1$, so if ε is a primitive nth root, then η is also.

If ε and ε' are any two roots of $\Phi_n(x)$, then the mapping $\varepsilon \to \varepsilon'$ determines an automorphism of the splitting field of $\Phi_n(x)$, because _____

_____.

Therefore, the group of $\Phi_n(x)$ is transitive and by the theorem quoted above, $\Phi_n(x)$ is irreducible. (For a different proof of this, see [vdW, p. 161 ; or MacD, p. 106].)

4.4 ROOTS OF UNITY AND CYCLIC EQUATIONS

(3) If ε and ε' are nth roots of unity and ε is primitive, then $\varepsilon' = \varepsilon^k$ for some k, $1 \leq k \leq n$. We shall show that ε' is primitive if and only if $\gcd(k, n) = 1$: If $\gcd(k, n) = a \neq 1$ and $k = ah$, $n = am$, then

$$(\varepsilon')^m = (\varepsilon^k)^m = \varepsilon^{(ah)m} = \varepsilon^{h(am)} = \varepsilon^{hn} = 1,$$

where $m < n$, because _____.

Therefore, ε' is not primitive. Therefore, if ε' is primitive, then $\gcd(k, n) = 1$.

If $\gcd(k, n) = 1$ and $(\varepsilon')^m = 1$, then $1 = (\varepsilon')^m = (\varepsilon^k)^m = \varepsilon^{km}$. Therefore, $km = ln$ for some integer l, so that $n|km$. But $\gcd(k, n) = 1$, so we have $n|m$. The element ε' must therefore be raised at least to the nth power before we get 1, so ε' is primitive.

There are, therefore, exactly as many primitive roots among $\varepsilon, \varepsilon^2, \ldots, \varepsilon^{n-1}$ as there are integers k less than n and prime to it, namely $\varphi(n)$. Each of these is a root of $\Phi_n(x)$ and each root of $\Phi_n(x)$ is of this form. Therefore, $\Phi_n(x)$ has exactly $\varphi(n)$ roots and must be of degree $\varphi(n)$. ‖

We now prove the theorem used in part (2):

THEOREM.   If $f(x) \in F[x]$ and $f(x)$ is separable and has no repeated factors, then

$$f(x) \text{ is irreducible} \Leftrightarrow \text{the Galois group of } f(x) \text{ is transitive.}$$

Proof. $\Rightarrow$ Let α, β be roots of $f(x)$ and let E be its splitting field. From Section 1.4 we know that since $f(x)$ is irreducible, the quotient $F[x]/f(x)$ is a field and

$$F(\alpha) \cong F[x]/(f(x)) \cong F(\beta).$$

Let σ be an isomorphism of $F(\alpha)$ onto $F(\beta)$. Since $F(\alpha)$ and $F(\beta)$ are both subfields of E and E is the splitting field of $f(x)$, this mapping can be extended to an automorphism of E. (See Section 3.1, p. 73.) Therefore, the Galois group of $f(x)$ contains an automorphism mapping α onto β for any two roots α and β. The group is therefore transitive.

$\Leftarrow$ We shall show that if $f(x)$ is separable and reducible, then its group is not transitive. For suppose $f(x) = g(x) \cdot h(x)$ and that $g(x)$ is irreducible. Since $f(x)$ is separable and has no repeated factors, no root of $g(x)$ is a root of $h(x)$ and g and h are relatively prime.

Let α be a root of $g(x)$, and let β be a root of $h(x)$. No automorphism σ of E can map α into β, for otherwise $0 = \sigma(0) = \sigma(g(\alpha)) = g(\sigma(\alpha)) = g(\beta)$, and this would make β a root of $g(x)$, which is impossible. Therefore G is not transitive. $\|$

We are now finally ready to prove the main theorem of this section.

THEOREM.   The cyclotomic polynomial $\Phi_n(x)$ and all cyclic equations are explicitly solvable in radicals over any field F for which we have a constructive factorization procedure.

Proof.   We shall prove this by induction on n. The induction hypothesis is (A) All cyclotomic polynomials $\Phi_k(x)$, and (B) all cyclic equations of degree k are explicitly solvable for all positive $k \le n - 1$. For $n = 1$ the result is clearly true.

Suppose the induction hypothesis holds. We must prove (A) that we can solve for all nth roots of unity, and (B) that we can solve all cyclic equations of degree n.

To prove (A): We have three cases:

(1) n is a prime p.

(2) n is a prime power p^c.

(3) n is a product of two or more powers of distinct primes $p_1^{c_1} \cdot p_2^{c_2} \cdot \ldots \cdot p_l^{c_l}$.

Case 1. If $n = p$, where p is a prime and ε is a primitive pth root of unity, then we know from Section 4.2 that the group $G(Q(\varepsilon)/Q)$ is isomorphic to $Z_p^\times$, which is cyclic and of order $p - 1$, that is, the nonzero integers modulo p under multiplication. Also

$$\Phi_p(x) = x^{p-1} + x^{p-2} + \cdots + x + 1.$$

Therefore, $\Phi_p(x)$ is a cyclic equation of degree $n - 1$ and solvable by the induction hypothesis.

4.4 ROOTS OF UNITY AND CYCLIC EQUATIONS

Case 2. If $n = p^c$ and ε is a p^cth root of unity, then ε is a primitive root of the equation

$$x^p = \zeta,$$

where ζ is a primitive p^{c-1}th root of unity. Its roots are $\varepsilon = \eta \sqrt[p]{\zeta}$, where η is a pth primitive root.

Case 3. Here $n = p_1^{c_1} \cdot p_2^{c_2} \cdot \ldots \cdot p_l^{c_l}$, with $l \geq 2$ and $p_i \neq p_j$ for $i \neq j$. We already know all primitive $p_i^{c_i}$th roots ζ_i in radicals, because _____

_____. The products

$$\zeta_1 \cdot \zeta_2 \cdot \ldots \cdot \zeta_l,$$

where ζ_i ranges over the primitive $p_i^{c_i}$th roots, will then range over all primitive nth roots, because _____.

These are precisely the roots of $\Phi_n(x)$, so case 3 holds.

To prove (B): We now use a procedure that was originally due to Lagrange and dates from 1770. Suppose the given cyclic equation is

$$x^n + a_{n-1}x^{n-1} + a_{n-2}x^{n-2} + \cdots + a_1 x + a_0 = 0,$$

with roots $\alpha_1, \ldots, \alpha_n$ numbered in such a manner that the permutation $\theta = (123 \cdots n)$ when applied to the subscripts of the α's generates the Galois group and that the coefficients $a_0, \ldots, a_{n-1}$ lie in the field F.

Form the expressions

$$
\begin{aligned}
f_1(x) &= \alpha_1 + x\alpha_2 + x^2\alpha_3 + \cdots + x^{n-2}\alpha_{n-1} + x^{n-1}\alpha_n, \\
f_2(x) &= \alpha_2 + x\alpha_3 + x^2\alpha_4 + \cdots + x^{n-2}\alpha_n \quad\;\; + x^{n-1}\alpha_1 \quad = \theta f_1(x), \\
f_3(x) &= \alpha_3 + x\alpha_4 + x^2\alpha_5 + \cdots + x^{n-2}\alpha_1 \quad\;\; + x^{n-1}\alpha_2 \quad = \theta^2 f_1(x), \\
&\;\;\vdots \\
f_n(x) &= \alpha_n + x\alpha_1 + x^2\alpha_2 + \cdots + x^{n-2}\alpha_{n-2} + x^{n-1}\alpha_{n-1} = \theta^{n-1}f_1(x).
\end{aligned}
$$

$$(1)$$

Let ζ be an nth root of unity, not necessarily primitive. Then $\zeta^n = 1$, and we see that

$$f_2(\zeta) = \zeta^{-1}f_1(\zeta),$$

$$f_3(\zeta) = \zeta^{-2}f_1(\zeta),$$

$$\vdots$$

$$f_n(\zeta) = \zeta^{-(n-1)}f_1(\zeta).$$

Raising each of these to the pth power gives

$$(f_1(\zeta))^p = (f_2(\zeta))^p = \cdots = (f_n(\zeta))^p,$$

and therefore there is some function $\psi(\zeta)$ such that $\psi(\zeta) = (f_n(\zeta))^p$ and

$$\theta\psi(\zeta) = \theta(f_i(\zeta))^p = (\theta^{i-1}f_1(\zeta))^p = (f_i(\zeta))^p = \psi(\zeta), \qquad \text{for all } i = 1, \ldots, n. \quad (2)$$

The last equations show that for any nth root ζ, the quantity $(f_i(\zeta))^p$ is invariant under the Galois group of the equation. It must therefore be an element of the field $F(\zeta)$, and once $\psi(\zeta)$ is known, we can evaluate $f_1(\zeta), \ldots, f_n(\zeta)$.

Moreover, if $f_1(\zeta), \ldots, f_n(\zeta)$ are known, we can evaluate $\alpha_1, \ldots, \alpha_n$ very simply as follows. First, remember that for any fixed integer k,

$$\sum_{\zeta^n = 1} \zeta^k = 0,$$

where the sum is taken over all distinct nth roots of unity. This holds because

_____. Therefore,

$$\sum_{\zeta^n = 1} f_1(\zeta) = \sum_{\zeta^n = 1} (\alpha_1 + \zeta\alpha_2 + \cdots + \zeta^{n-1}\alpha_n)$$

$$= n\alpha_1 + \Sigma\zeta\alpha_2 + \cdots + \Sigma\zeta^{n-1}\alpha_n$$

$$= n\alpha_1 + \alpha_2\Sigma\zeta + \cdots + \alpha_n\Sigma\zeta^{n-1}$$

$$= n\alpha_1,$$

and, in general,

$$\sum_{\zeta^n=1} f_i(\zeta) = \sum_{\zeta^n=1} (\alpha_i + \zeta\alpha_{i+1} + \cdots + \zeta^{n-1}\alpha_{i-1})$$

$$= n\alpha_i,$$

so we can calculate $\alpha_1, \ldots, \alpha_n$ very easily once we know the values of $f_i(\zeta)$. These remain to be evaluated and the last step now must show how to do this, that is, how to evaluate $\psi(\zeta)$. From equations (1) and (2) it is evident that $\psi(\zeta)$ depends on $\alpha_1, \ldots, \alpha_n$ and we know that it is invariant under the group G of the given equation. We have $G < \mathfrak{S}_n$ and the numerical value of $\psi(\zeta)$ for any given ζ must also be invariant under $\mathfrak{S}_n$, since it is an element of $F(\zeta)$, but as a formal expression it is not necessarily invariant under $\mathfrak{S}_n$.

For example, suppose $\alpha_1, \alpha_2, \alpha_3$ are the roots of the cyclic equation $x^3 - 3x + 1 = 0$. (This equation will be discussed in greater detail in Section 4.5.) Then $n = 3$, so that the possible values of ζ are 1, ω, or ω^2 and

$$f_1(\zeta) = \alpha_1 + \alpha_2\zeta + \alpha_3\zeta^2,$$

$$\psi(\zeta) = (f_1(\zeta))^3 = \alpha_1^3 + \alpha_2^3 + \alpha_3^3 + 3\zeta(\alpha_1\alpha_3^2 + \alpha_3\alpha_2^2 + \alpha_2\alpha_1^2)$$

$$+ 3\zeta^2(\alpha_1\alpha_2^2 + \alpha_2\alpha_3^2 + \alpha_3\alpha_1^2) + 6\alpha_1\alpha_2\alpha_3,$$

and we shall see that for this equation $\psi(1) = 0$, $\psi(\omega) = -27\omega$, and $\psi(\omega^2) = -27\omega^2$. (Our ψ here is the φ_0 of Section 4.5, p. 176.) Therefore,

$$\psi(\omega) = -27\omega = \alpha_1^3 + \alpha_2^3 + \cdots + 6\alpha_1\alpha_2\alpha_3,$$

and the expression on the left side of the equation is invariant under $\mathfrak{S}_n$, but the one on the right side is not. If it were, it would be a symmetric function of $\alpha_1, \alpha_2, \alpha_3$ and we could evaluate it from the coefficients of the given equation—these coefficients have not been used at all so far.

To find the numerical value of $\psi(\zeta)$ we can proceed as follows. Let the above expression for $\psi(\zeta)$ in terms of $\alpha_1, \ldots, \alpha_n$ be denoted by ψ_1 and form the expressions $\psi_1, \psi_2, \ldots, \psi_s$ which are the distinct conjugates of ψ_1 under the permutations of $\mathfrak{S}_n$. In the example above we would have

$$\psi(\zeta) = \psi_1 = (1)\psi_1 = (123)\psi_1 = (132)\psi_1,$$

$$\psi_2 = (12)\psi_1 = (23)\psi_1 = (13)\psi_1 = $$

_____ in terms of

$\alpha_1, \alpha_2, \alpha_3$. Then the coefficients of the equation

$$(Y - \psi_1)(Y - \psi_2) \cdots (Y - \psi_s) = 0 \tag{3}$$

are all symmetric functions of $\psi_1, \ldots, \psi_s$, and thus symmetric in $\alpha_1, \ldots, \alpha_n$. They can therefore be evaluated from the coefficients $a_0, \ldots, a_{n-1}$ of the given equation. (This is where the coefficients are finally used.) So we can rewrite equation (3) as

$$Y^s + b_{s-1}Y^{s-1} + \cdots + b_0 = 0 \tag{4}$$

with $b_0, \ldots, b_{s-1}$ known elements of F. However, our hypothesis was that the group of the equation was cyclic and we already saw that ψ_1 would therefore be an element of F. So equation (4) must factor in F with at least one linear factor. If we have a constructive method of factoring in F, we can then solve (4) and find the value of ψ_1. In the field of rationals Q and in all finite algebraic extensions of Q there is a constructive factorization procedure due to Kronecker [vdW, p. 77], so the procedure for a solution of cyclic equations can always be carried out. The additional hypothesis on F introduced here—that there be a constructive factorization procedure—is necessary if we wish to show that there is an explicit solution in radicals. If we only wish to prove the existence of a solution in radicals, it is not necessary.

We have already shown how to obtain $\alpha_1, \ldots, \alpha_n$ from ψ_1, so this concludes the proof of the theorem. ‖

4.4 ROOTS OF UNITY AND CYCLIC EQUATIONS

As examples we shall evaluate the fifth and seventh roots of unity. In Section 4.6 we shall evaluate the seventeenth roots and construct a regular polygon of 17 sides. There is also an additional example of a cyclic equation in Section 4.5.

EXAMPLE (1) *The fifth roots of unity.* We must solve the equation $x^5 - 1 = 0$. Factoring gives

$$x^5 - 1 = (x - 1)(x^4 + x^3 + x^2 + x + 1),$$

and so

$$\Phi_5(x) = x^4 + x^3 + x^2 + x + 1.$$

This equation must be cyclic, because _____

_____, but instead of applying the method for cyclic equations of degree 4 directly, it will save a great deal of work if we first reduce the degree of the equation by the following trick: If x is a root of $\Phi_5(x)$, then $x \neq 0$, because _____

_____, so we can divide through by x^2 and get the equation

$$x^2 + x + 1 + \frac{1}{x} + \frac{1}{x^2} = 0. \tag{5}$$

Let $y = x + 1/x$, so that $y^2 = $ _____. Equation (5) then becomes

$$\underline{\hspace{10cm}},$$

whose roots are

$$y = \underline{\hspace{6cm}} = x + \frac{1}{x}.$$

To find x, we must therefore still solve the equation

$$x + \frac{1}{x} = y \qquad \text{or, equivalently,} \qquad x^2 - xy + 1 = 0.$$

The roots of this equation are

$$x = \underline{\hspace{4cm}} = \tfrac{1}{4}(-1 \pm \sqrt{5}) \pm \frac{i}{2}\sqrt{\frac{5 \pm \sqrt{5}}{2}}, \quad (6)$$

where the signs in front of $\sqrt{5}$ must be the same and we can choose either sign in front of $i/2$, because \underline{\hspace{7cm}}.

Using equation (6) it is very easy to construct a regular pentagon by ruler and compass, if we remember that the various nth roots of unity lie on the vertices of a regular n-gon inscribed in the unit circle, because a primitive nth root of unity corresponds to the point with polar coordinates (\underline{\hspace{1cm}}, \underline{\hspace{1cm}}) in the complex plane. For $n = 5$, we need only construct the root

$$\varepsilon = \tfrac{1}{4}(-1 + \sqrt{5}) + \frac{i}{2}\sqrt{\frac{5 + \sqrt{5}}{2}}.$$

Construction of a Regular Pentagon

(1) Fix the scale by deciding on a unit length, say 1 unit $= 2$ inches.

(2) Draw a right triangle with legs of length 1 and 2 units.

(3) The length of the hypotenuse will then be \underline{\hspace{1.5cm}} units.

(4) Construct a segment of length $(\sqrt{5} - 1)$ units.

(5) Construct a segment of length $\tfrac{1}{4}(\sqrt{5} - 1)$ units.

(6) Draw a circle with radius 1 unit.

(7) Taking the center of this circle as the origin, locate the point A on the x axis whose x coordinate is $\tfrac{1}{4}(\sqrt{5} - 1)$.

(8) Draw a vertical line through A.

(9) The intersection of this line and the circle will then be ε, because \underline{\hspace{3cm}}

\underline{\hspace{8cm}}.

(10) Complete the pentagon.

150

4.4 ROOTS OF UNITY AND CYCLIC EQUATIONS

EXAMPLE (2)   *The seventh roots of unity.* We must solve the equation $x^7 - 1 = 0$. Factoring gives

$$x^7 - 1 = (x - 1)(x^6 + x^5 + x^4 + x^3 + x^2 + x + 1) = 0,$$

and so

$$\Phi_7(x) = x^6 + x^5 + x^4 + x^3 + x^2 + x + 1.$$

We can again halve the degree of this equation, this time by dividing by x^3 before substituting $y = x + (1/x)$. (This is a trick that works for all "reciprocal" equations. They are equations of the form $a_n x^n + a_{n-1}x^{n-1} + a_{n-2}x^{n-2} + \cdots + a_{n-2}x^2 + a_{n-1}x + a_n = 0$. This can easily be proved by induction on n.)

We have

$$y = x + \frac{1}{x}, \quad y^2 = \underline{\hspace{3cm}}, \quad y^3 = \underline{\hspace{3cm}},$$

and the equation

$$x^3 + x^2 + x + 1 + \frac{1}{x} + \frac{1}{x^2} + \frac{1}{x^3} = 0. \tag{7}$$

On carrying out the substitution we get

$$y^3 + y^2 - 2y - 1 = 0. \tag{8}$$

This equation must be cyclic because $\underline{\hspace{5cm}}$.

To solve equation (8) by the general method for cyclic equations, let $\alpha_1, \alpha_2, \alpha_3$ be the roots. The Galois group of (8) is

$$G = \{(1), (123), (132)\} \quad \text{and} \quad \theta = (123).$$

Then

$$f_1(x) = \alpha_1 + x\alpha_2 + x^2\alpha_3,$$

$$f_2(x) = \alpha_2 + x\alpha_3 + x^2\alpha_1,$$

$$f_3(x) = \alpha_3 + x\alpha_1 + x^2\alpha_2,$$

and for the cube roots of unity $\zeta = 1$, ω, or ω^2 we get

$$\psi(1) = (f_1(1))^3 = (f_2(1))^3 = (f_3(1))^3 = (\alpha_1 + \alpha_2 + \alpha_3)^3,$$

$$\psi(\omega) = (f_1(\omega))^3 = (f_2(\omega))^3 = (f_3(\omega))^3 = (\alpha_1 + \omega\alpha_2 + \omega^2\alpha_3)^3,$$

$$\psi(\omega^2) = (f_1(\omega^2))^3 = (f_2(\omega^2))^3 = (f_3(\omega^2))^3 = (\alpha_1 + \omega^2\alpha_2 + \omega^4\alpha_3)^3.$$

From the coefficients of equation (8), we get

$$\sum \alpha_i = -1, \qquad \sum_{i \neq j} \alpha_i\alpha_j = -2, \qquad \sum_{i \neq j \neq k \neq i} \alpha_i\alpha_j\alpha_k = \alpha_1\alpha_2\alpha_3 = 1.$$

We therefore have immediately that $f_1(1) = f_2(1) = f_3(1) = -1$ and $\psi(1) = (-1)^3 = -1$. More work is necessary to evaluate $\psi(\omega)$ and $\psi(\omega^2)$. We get

$$\psi(\omega) = (\alpha_1^3 + \alpha_2^3 + \alpha_3^3) + 3\omega(\alpha_1\alpha_3^2 + \alpha_3\alpha_2^2 + \alpha_2\alpha_1^2)$$

$$+ 3\omega^2(\alpha_1\alpha_2^2 + \alpha_2\alpha_3^2 + \alpha_3\alpha_1^2) + 6\alpha_1\alpha_2\alpha_3.$$

Let

$$A = \alpha_1\alpha_3^2 + \alpha_3\alpha_2^2 + \alpha_2\alpha_1^2,$$

$$B = \alpha_1\alpha_2^2 + \alpha_2\alpha_3^2 + \alpha_3\alpha_1^2.$$

(This same notation and some of the same calculations will appear in the next section.)
We then have

$$\psi(\omega) = \sum \alpha_i^3 + 3\omega A + 3\omega^2 B + 6\alpha_1\alpha_2\alpha_3,$$

$$\psi(\omega^2) = \underline{\hspace{6cm}}.$$

4.4 ROOTS OF UNITY AND CYCLIC EQUATIONS

Using the theory of symmetric functions we see that

$$\sum \alpha_i^3 = (\sum \alpha_i)^3 - 3(\sum \alpha_i)(\sum \alpha_i \alpha_j) + 3 \sum \alpha_i \alpha_j \alpha_k$$

$$= \underline{\hspace{2cm}} - 3 \cdot \underline{\hspace{2cm}} \cdot \underline{\hspace{2cm}} + 3 \cdot \underline{\hspace{2cm}} = \underline{\hspace{1.5cm}}$$

$$($$

when we substitute in the numerical values obtained from the coefficients of equatio
(8). The first line of equation (9) can be checked by multiplying out.

However, A and B are not symmetric functions of $\alpha_1, \alpha_2, \alpha_3$. They are invarian
under the cyclic group G, so the numerical values of A and B must be elements of $Q(\omega)$
To find these values, we must find all their conjugates under $\mathfrak{S}_3$. We get

$$(1)A = (123)A = (132)A = A,$$

$$(12)A = (23)A = (13)A = \underline{\hspace{1.5cm}},$$

$$(1)B = (123)B = (132)B = B,$$

$$(12)B = (23)B = (13)B = \underline{\hspace{1.5cm}}.$$

Therefore, the equation

$$(z - A)(z - B) = 0$$

must have roots in $Q(\omega)$.

Since

$$(z - A)(z - B) = z^2 - (A + B)z + AB, \tag{10}$$

we first evaluate $A + B$ and AB. Both are symmetric and we have

$$A + B = \sum_{i \neq j} \alpha_i \alpha_j^2 = -3\sum \alpha_i \alpha_j \alpha_k + (\sum \alpha_i) \cdot (\sum \alpha_i \alpha_j)$$

$$= -3 \cdot (\underline{\hspace{1cm}}) + (\underline{\hspace{1cm}}) \cdot (\underline{\hspace{1cm}}) = \underline{\hspace{1cm}},$$

$AB =$ _____

_____

$$= 3\alpha_1^2\alpha_2^2\alpha_3^2 + \sum \alpha_i\alpha_j\alpha_k^4 + \sum \alpha_i^3\alpha_j^3$$

$$= 9(\alpha_1\alpha_2\alpha_3)^2 + \left(\sum \alpha_i\right)^3 \cdot (\alpha_1\alpha_2\alpha_3) + \left(\sum \alpha_i\alpha_j\right)^3$$

$$- 6\left(\sum \alpha_i\right)\left(\sum \alpha_i\alpha_j\right) \cdot (\alpha_1\alpha_2\alpha_3)$$

$$= 9(\underline{\quad})^2 + (\underline{\quad})^3(\underline{\quad}) + (\underline{\quad})^3$$

$$- 6(\underline{\quad}) \cdot (\underline{\quad}) \cdot (\underline{\quad})$$

$$= \underline{\qquad}.$$

The above calculations involving the symmetric functions should be checked rather carefully. Substituting these values into equation (10) we find that we must solve the equation

$$z^2 + z - 12 = 0.$$

The solutions are $z = +3, -4$. Which of the two solutions is A and which B depends on the way the roots are numbered. Suppose $A = 3$ and $B = -4$. Substituting these values and using equation (9) gives

$$\psi(\omega) = -4 + 9\omega - 12\omega^2 + 6$$

$$= \tfrac{7}{2}(1 + 3\sqrt{-3}),$$

$$\psi(\omega^2) = -4 - 12\omega + 9\omega^2 + 6$$

$$= \tfrac{7}{2}(1 - 3\sqrt{-3}).$$

4.4 ROOTS OF UNITY AND CYCLIC EQUATIONS

Therefore,

$$f_1(\omega) = \sqrt[3]{\psi(\omega)} \qquad = \sqrt[3]{\tfrac{7}{2}(1 + 3\sqrt{-3})},$$

$$f_2(\omega) = \underline{\hspace{2cm}} = \omega\sqrt[3]{\tfrac{7}{2}(1 + 3\sqrt{-3})}, \qquad (11)$$

$$f_3(\omega) = \underline{\hspace{2cm}} = \omega^2\sqrt[3]{\tfrac{7}{2}(1 + 3\sqrt{-3})}.$$

$$f_1(\omega^2) = \qquad \sqrt[3]{\psi(\omega^2)} = \sqrt[3]{\tfrac{7}{2}(1 - 3\sqrt{-3})},$$

$$f_2(\omega^2) = \underline{\hspace{2cm}} = \begin{pmatrix} \omega \\ \text{or} \\ \omega^2 \end{pmatrix} \cdot \sqrt[3]{\tfrac{7}{2}(1 - 3\sqrt{-3})}, \qquad (12)$$

$$f_3(\omega^2) = \underline{\hspace{2cm}} = \begin{pmatrix} \omega \\ \text{or} \\ \omega^2 \end{pmatrix} \cdot \sqrt[3]{\tfrac{7}{2}(1 - 3\sqrt{3})},$$

because we know that $f_1(\omega) \neq f_2(\omega) \neq f_3(\omega)$, for if two of them were the same, say if $f_1(\omega) = f_2(\omega)$, then $\underline{\hspace{5cm}}$

$\underline{\hspace{8cm}}$.

It is not clearly specified which cube root is meant by $\sqrt[3]{\tfrac{7}{2}(1 \pm 3\sqrt{-3})}$, as this depends on the numbering of the roots, but it does not matter. However in equations (12), the choice of factor for $f_2(\omega^2)$ and $f_3(\omega^2)$ does matter and will have to be adjusted to make the roots satisfy equation (8). Remembering that $f_1(1) = f_2(1) = \alpha_1 + \alpha_2 + \alpha_3 = -1$, we have the solutions

$$\alpha_1 = \tfrac{1}{3}(f_1(1) + f_1(\omega) + f_1(\omega^2))$$

$$= \tfrac{1}{3}(-1 + \sqrt[3]{\tfrac{7}{2}(1 + 3\sqrt{-3})} + \sqrt[3]{\tfrac{7}{2}(1 - 3\sqrt{-3})}),$$

$$\alpha_2 = \tfrac{1}{3}(f_2(1) + f_2(\omega) + f_2(\omega^2))$$

$$= \tfrac{1}{3}(-1 + \omega^2 \sqrt[3]{\tfrac{7}{2}(1 + 3\sqrt{-3})} + \omega \sqrt[3]{\tfrac{7}{2}(1 - 3\sqrt{-3})}),$$

$$\alpha_3 = \tfrac{1}{3}(-1 + \omega \sqrt[3]{\tfrac{7}{2}(1 + 3\sqrt{-3})} + \omega^2 \sqrt[3]{\tfrac{7}{2}(-3\sqrt{-3})}),$$

where we have chosen the factors ω and ω^2 for $f_2(\omega^2)$ and $f_3(\omega^2)$, respectively. This choice is only justified by the fact that substituting these values back into equation (8) gives a solution, whereas the alternative choices will not.

We now have the three values α_1, α_2, α_3 of y, and must still solve the equations

$$x + \frac{1}{x} = \alpha_i,$$

for $i = 1, 2, 3$, to get the seventh roots of unity. The solutions are

$$x = \tfrac{1}{2}(\alpha_i \pm \sqrt{\alpha_i^2 - 4}), \qquad i = 1, 2, 3.$$

There is a famous method of evaluating the pth roots of unity due to Gauss (see the end of Section 4.6 or [vdW, p. 163]), which is essentially an application of the general process for solvable equations to this case.

In closing this section, we list in Table 4.1 a few of the primitive roots of unity expressed in radicals.

TABLE 4.1   Table of Primitive nth Roots of Unity

n	ε_n
1	1
2	$\varepsilon_2 =$

4.4 ROOTS OF UNITY AND CYCLIC EQUATIONS

TABLE 4.1 Table of Primitive nth Roots of Unity (contd.)

n	ε_n
3	$\varepsilon_3 = \omega = -\frac{1}{2} + \frac{1}{2}\sqrt{-3}$
4	$\varepsilon_4 = i$
5	$\varepsilon_5 =$
6	$\varepsilon_6 = -\omega =$
7	$\varepsilon_7 =$
8	$\varepsilon_8 = \sqrt{i} = \frac{1}{2}(\sqrt{2} + i\sqrt{2})$
10	$\varepsilon_{10} = -\varepsilon_5 =$
12	$\varepsilon_{12} = \varepsilon_3\varepsilon_4 =$
	From Section 4.6:
17	$\varepsilon_{17} =$

4.5 HOW TO SOLVE A SOLVABLE EQUATION

We saw that solvable equations have solvable groups and so, because there are indeed some equations whose groups are not solvable, we know that these equations are not solvable in radicals. They are solvable by other means. For example, there are matrix solutions and approximate solutions, but at the moment we are not considering these. (Matrix solutions are discussed in Section 4.10; the best known approximation method is Newton's method.)

If the group G of the equation $p(x) = 0$ is solvable, then it is quite easy to prove the existence of a solution in radicals.

THEOREM.   If G is the group of $p(x)$ over F and G is solvable, then $p(x)$ has a solution in radicals. (We assume that char $F = 0$.)

Proof.   Corresponding to the composition series

$$G = G_0 \underset{p_1}{\rhd} G_1 \underset{p_2}{\rhd} \cdots \underset{p_r}{\rhd} G_r = (e)$$

of the solvable group G with G_i of prime index p_i in G_{i-1}, we have a sequence of fields

$$F = F_0 \lhd F_1 \lhd \cdots \lhd F_r = E,$$

where $G(F_i/F_{i-1}) = G_{i-1}/G_i$ and F_i is therefore a cyclic extension of order p_i. From Section 4.4. we know that F_i can then be obtained from F_{i-1} by adjoining ε_{p_i} and elements of the form $\sqrt[p_i]{\gamma}$ with $\gamma \in F_{i-1}$. The splitting field E must then be an extension of F by radicals and so $p(x)$ has a solution in radicals.   ‖

However, merely knowing that there is some radical extension which contains the roots and actually exhibiting them and writing them down are two different matters, and we now show how to find the roots.

So suppose now that we have in front of us a polynomial $p(x) = x^n + a_{n-1}x^{n-1} + \cdots + a_0$ together with its group G and a composition series for G in which all the composition factors are cyclic of prime order. How shall we go about the solution of this equation? Where does one start? In this section we shall see how to find the roots of $p(x)$ in radicals whenever this is possible at all. The reduction step used in the procedure described here is essentially Lagrange's solution of cyclic equations. It makes use of various nth roots of unity, but at least in fields of characteristic 0 we now know that these can always be expressed in terms of radicals, as in Section 4.4 or by using Gauss's method of solution [vdW, p. 163]. Historically, however, Lagrange's solution of cyclic equations precedes Gauss's evaluation of the roots of unity. The content of this section can be summarized in the form of a theorem:

THEOREM.   Let F be a field, and let $p(x) \in F[x]$ be of degree n. Suppose we can form F_ε by adjoining to F a primitive kth root of unity for every k such that $k|n$ and also that the Galois group G of $p(x)$ over F is solvable. If, in addition, we have a factorization algorithm for $F_\varepsilon[x]$, then there is an algorithm for finding the roots of $p(x)$ in terms of radicals.

Instead of a formal proof of this theorem there will be a description of the method of solution interspersed with examples. A formal proof could easily be constructed from this by eliminating the examples and some of the explanation and motivation.

It is essential here that the Galois group of $p(x)$ be expressed as a permutation group of the roots of $p(x)$. It is not sufficient to know G only as an abstract group or to within isomorphism. This is a reasonable condition, because, for example, the polynomials

$$(x^2 + 1)(x^2 + 2)$$

and

$$x^4 - 2x^2 + 9$$

have the same splitting field and therefore the same abstract group, but you would not use the same procedure to solve the first as the second, and you will see that the permutations in the group determine the process of solution.

So suppose now that the composition series of the Galois group of the given nth-degree polynomial $p(x) \in F[x]$ is

$$G = G_0 \underset{p_1}{\triangleright} G_1 \underset{p_2}{\triangleright} G_2 \underset{p_3}{\triangleright} G_3 \underset{p_4}{\triangleright} \cdots \underset{p_r}{\triangleright} G_r = (1),$$

where p_i is the index of G_i in G_{i-1} and p_i is a prime. If we let $H_i = G_{i-1}/G_i$, then the group H_i is of order p_i and so is cyclic (all groups of prime order are cyclic), making G solvable. The roots of $p(x)$ will be called $\alpha_1, \alpha_2, \ldots, \alpha_n$. The group G is assumed to be expressed as a permutation group of the integers $1, 2, \ldots, n$ and so it is a sub-group of $\mathfrak{S}_n$. Throughout this section ζ_i will be a p_ith root of unity.

For example, if $G_1 = \mathfrak{S}_4$ we have a composition series for which

$$r = 4, G = G_0 = \mathfrak{S}_4, G_1 = \mathfrak{A}_4, G_2 = \mathfrak{B}, G_3 = \mathfrak{H}, G_4 = (1),$$

where, of course,

$$\mathfrak{S}_4 \underset{p_1=2}{\triangleright} \mathfrak{A}_4 \underset{p_2=3}{\triangleright} \mathfrak{B} \underset{p_3=2}{\triangleright} \mathfrak{H} \underset{p_4=2}{\triangleright} (1).$$

Here $\mathfrak{A}_4$ is again the alternating group,

$$\mathfrak{S}_4 \underset{p_1=2}{\triangleright} \mathfrak{A}_4 \underset{p_2=3}{\triangleright} \mathfrak{B} \underset{p_3=2}{\triangleright} \mathfrak{H} \underset{p_4=2}{\triangleright} (1).$$

$$\mathfrak{H} = \{(1), (12)(34)\}.$$

Let the brackets $[a]$ denote the coset in G_{i-1} of the enclosed element a of G_i. The quotient groups then are

$$H_1 = G_0/G_1 = \mathfrak{S}_4/\mathfrak{A}_4 = \{[(1)], [(12)]\} = \{\mathfrak{A}_4, [(12)]\}$$

$$= \{\{(1), (12)(34), (13)(24), (14)(23), (123), (132), (124), (142),$$

$$(134), (143), (234), (243)\},$$

$$\{(12), (13), (14), (23), (24), (34), (1234), (1324), (1423), (1432),$$

$$(1243), (1342)\}\},$$

4.5 HOW TO SOLVE A SOLVABLE EQUATION

$$H_2 = G_1/G_2 = \mathfrak{A}_4/\mathfrak{B} = \{[(1)], [(123)], [132]\} = \{G_2, [(123)], [132]\}$$

$$= \{\{(1), (12)(34), (13)(24), (14)(23)\}, \{(123), \underline{\hspace{2cm}}, \underline{\hspace{2cm}},$$

$$\underline{\hspace{2cm}}\},$$

$$\{(132), \underline{\hspace{2cm}}, \underline{\hspace{2cm}}, \underline{\hspace{2cm}}\}\},$$

$$H_3 = G_2/G_3 = \mathfrak{B}/\mathfrak{H} = \{[(1)], [(13)(24)]\} = \{G_3, [(13)(24)]\}$$

$$= \{\{\underline{\hspace{2cm}}, \underline{\hspace{2cm}}\}, \{\underline{\hspace{2cm}}, \underline{\hspace{2cm}}\}\},$$

$$H_4 = G_3/G_4 = \mathfrak{H}/(1) = \{\{(1)\}, \{(12)(34)\}\}.$$

We shall use the composition series to construct functions $\{\varphi_{i-1}(\zeta_i) \mid 1 \le i \le r,$ $\zeta_i^{p_i} = 1\}$ of $\alpha_1, \ldots, \alpha_n$ and a p_ith root ζ_i whose numerical value depends, of course, indirectly on the coefficients of the given polynomial $p(x)$ and which we shall be able to evaluate numerically to yield expressions for the roots in radicals. These functions will satisfy the following conditions:

(1) For every i, each $\varphi_i(\zeta_{i+1})$ is invariant under all of the permutations of G_i.

(2) The numerical value of the functions $\varphi_0(\zeta_1)$ can be obtained from the coefficients of $p(x)$ using only the usual rational field operations $+, -, \times, \div$. Their values are elements of F.

(3) For $i \ge 2$, each of the functions $\varphi_i(\zeta_{i+1})$ can be evaluated in terms of $\{\varphi_{i-1}(\zeta_i) \mid \zeta_i^{p_i} = 1\}$ using only the operations $+, -, \times, \div$, and extraction of p_ith roots of elements of $F_\varepsilon\{\varphi_{i-1}(\zeta_i) \mid \zeta_i^{p_i} = 1\}$.

(4) The roots $\alpha_1, \ldots, \alpha_n$ of $p(x)$ can be evaluated in terms of $\{\varphi_{r-1}(\zeta_r) \mid \zeta_r^{p_r} = 1\}$ using only the operations $+, -, \times, \div$, and extraction of p_rth roots of elements of $F_\varepsilon\{\varphi_{r-1}\}$.

Functions that satisfy these conditions can generally be chosen in a number of different ways. The method here will not always result in the most convenient possible such set (in fact, it does not even do so for the general quartic); on the other hand, it does describe one such set for every solvable group G, so at least one can be sure of their existence and constructibility.

In the construction of these functions and in carrying out the solution, remember that F_ε is assumed to contain all necessary kth roots of unity already. The procedure we follow consists essentially of repeated applications of Lagrange's procedure for cyclic equations. This is natural, because all the quotient groups are cyclic so that F_{G_j} is always a cyclic extension of $F_{G_{j-1}}$.

As we do not yet know the values of the roots, it is surely no restriction to assume that they are numbered in such a way that

$$G_{r-1} \cong \{(1), \theta, \theta^2, \ldots, \theta_r^{p-1}\} = C_{p_r},$$

where θ is the permutation $(12 \ldots p_r)$ (θ is applied to the subscripts of the α's). In general, we cannot be sure that $G_{r-1} = G_{p_r}$, as we see easily by examining H_4 in the example. Since p_r is a prime, we can, however, be sure that G_{r-1} is generated by a permutation θ' which is of the form

$$(12 \ldots p_r)((p_r + 1)(p_r + 2) \ldots (2p_r))((2p_r + 1)(2p_r + 2) \ldots 3p_r)) \ldots$$

$$((k-1)(p_r + 1) \ldots kp_r),$$

that is, a product of disjoint p_r cycles, where, with appropriate numbering, the first one is just θ.

We now form the expressions

$$f_{r,1}(x) = \alpha_1 + x\alpha_2 + \cdots + x^{p_r-2}\alpha_{p_r-1} + x^{p_r-1}\alpha_{p_r},$$

$$f_{r,2}(x) = \alpha_2 + x\alpha_3 + \cdots + x^{p_r-2}\alpha_{p_r} + x^{p_r-1}\alpha_1 = \theta f_{r,1}(x),$$

$$f_{r,3}(x) = \alpha_3 + x\alpha_4 + \cdots + x^{p_r-2}\alpha_1 + x^{p_r-1}\alpha_2 = \theta^2 f_{r,1}(x),$$

$$\vdots$$

$$f_{r,p_r}(x) = \alpha_{p_r} + x\alpha_1 + \cdots + x^{p_r-2}\alpha_{p_r-2} + x^{p_r-1}\alpha_{p_r-1} = \theta^{p_r-1} f_{r,1}(x),$$

where θ, of course, is applied to the subscripts of the α's.

4.5 HOW TO SOLVE A SOLVABLE EQUATION

As in Section 4.4 we let $x = \zeta_r$, where now ζ_r is a p_rth root of unity, to form all possible functions $f_{r,j}(\zeta_r)$. Since $\zeta_r^{p_r} = 1$, we see that

$$f_{r,2}(\zeta_r) = \zeta_r^{-1} f_{r,1}(\zeta_r), \qquad f_{r,3}(\zeta_r) = \zeta_r^{-2} f_{r,1}(\zeta_r),$$

and so on, so that

$$(f_{r,1}(\zeta_r))^{p_r} = \cdots = (f_{r,p_r}(\zeta_r))^{p_r}.$$

Now let

$$\varphi_{r-1}(\zeta_r) = (f_{r,1}(\zeta_r))^{p_r}.$$

(Of course, φ_{r-1} depends also on $\alpha_1, \ldots, \alpha_n$, not only on ζ_r.)

As the construction and description of the general case are fairly intricate, let us now carry out the part described so far on a specific example, a general quartic. Suppose $F = Q$, $n = 4$, and

$$p(x) = c_0 + c_1 x + c_2 x^2 + c_3 x^3 + c_4 x^4,$$

where the group of $p(x)$ is $\mathfrak{S}_4$. From page 159 we see that $r = \underline{\hspace{1cm}}$, $p_r = \underline{\hspace{1cm}}$, and we form first

$$f_{4,1}(x) = \alpha_1 + x\alpha_2,$$

$$f_{4,2}(x) = \alpha_2 + x\alpha_1 = (12)(34)f_{4,1}(x) = (12)f_{4,1}(x),$$

because here $\theta' = (12)(34)$ and $\theta = (12)$. The only square roots of unity are $\zeta_4 = +1$ or -1, so we get

$$f_{4,1}(1) = \alpha_1 + \alpha_2,$$

$$f_{4,2}(1) = \alpha_2 + \alpha_1 = f_{4,1}(1),$$

$$f_{4,1}(-1) = \alpha_1 - \alpha_2,$$

$$f_{4,2}(-1) = \alpha_2 - \alpha_1 = (-1)f_{4,1}(-1).$$

Then we have

$$\varphi_3(1) = (f_{4,1}(1))^2 \quad = (f_{4,2}(1))^2 \quad = (\alpha_1 + \alpha_2)^2 = \alpha_1^2 + 2\alpha_1\alpha_2 + \alpha_2^2,$$

$$\varphi_3(-1) = (f_{4,1}(-1))^2 = (f_{4,2}(-1))^2 = (\alpha_1 - \alpha_2)^2 = \alpha_1^2 - 2\alpha_1\alpha_2 + \alpha_2^2,$$

so we clearly have each φ_3 invariant under G_3, that is, under $\mathfrak{H}$. Moreover,

$$\sum_{\zeta_4} f_{4,1}(\zeta_4) = f_{4,1}(1) + f_{4,1}(-1) = (\alpha_1 + \alpha_2) + (\alpha_1 - \alpha_2)$$

$$= 2\alpha_1,$$

$$\sum_{\zeta_4} f_{4,2}(\zeta_4) = f_{4,2}(1) + f_{4,2}(-1) = (\alpha_2 + \alpha_1) + (\alpha_2 - \alpha_1)$$

$$= 2\alpha_2,$$

which shows that α_1 and α_2 can be evaluated from the functions $f_{4,j}(\zeta_4)$, and these in turn are obtained from the $\varphi_3(\zeta_4)$ by extraction of square roots. This situation already illustrates the general case, to which we now return.

As we saw before, $\varphi_{r-1}(\zeta_r)$ is clearly unchanged by any power of θ or θ' and so is invariant under G_{r-1}. Therefore, $\varphi_{r-1} \in F_{G_{r-1}}$ and each $f_{r,j}(\zeta_r)$ is a p_rth root of an element of $F_{G_{r-1}}$. If each $\varphi_{r-1}(\zeta_r)$ is expressed in terms of radicals, then each $f_{r,j}(\zeta_r)$ can be, too.

Now summing $f_{r,j}(\zeta_r)$ as ζ_r varies over all p_rth roots of unity we get

$$\sum_{\zeta_r} f_{r,1}(\zeta_r) = p_r\alpha_1 + \left(\sum_{\zeta_r} \zeta_r\right)\alpha_2 + \cdots + \left(\sum_{\zeta_r} \zeta_r^{(p_r-1)}\right)\alpha_{p_r}$$

$$= p_r\alpha_1, \qquad \alpha_1 = \underline{},$$

because $\sum_{\zeta_r} \zeta_r = 0$ and the sums all extend over all p_rth roots of unity of which there are exactly p_r.

To solve for $\alpha_2, \alpha_3, \ldots, \alpha_{p_r}$ we see that

$$f_{r,2}(\zeta_r) = \zeta_r^{-1} f_{r,1}(\zeta_r) = \zeta_r^{-1}\alpha_1 + \alpha_2 + \zeta_r\alpha_2 + \cdots + \zeta_r^{(p_r-2)}\alpha_{p_r},$$

4.5 HOW TO SOLVE A SOLVABLE EQUATION

and summing again over all p_rth roots of unity we get

$$\sum_{\zeta_r} f_{r,2}(\zeta_r) = p_r\alpha_2, \qquad \alpha_2 = \underline{\hspace{3cm}}.$$

Similarly,

$$\sum_{\zeta_r} f_{r,p_r}(\zeta_r) = p_r\alpha_{p_r}, \qquad \alpha_{p_r} = \underline{\hspace{3cm}},$$

so the roots $\alpha_1, \ldots, \alpha_{p_r}$ can be obtained on division by p_r. If we assume that $p_r \nmid \text{char } F$ (or, since char F must be a prime, that $p_r \neq \text{char } F$), this division is possible. Two questions arise immediately:

(1) How do we evaluate $\varphi_{r-1}(\zeta_r)$?

(2) If $p_r \neq n$, how do we find the other roots $\alpha_{p_r+1}, \ldots, \alpha_n$?

First we turn to question (2): We shall find that since the functions $f_{r,j}$ are p_rth roots of (presumably) previously determined functions φ_{r-1} they may be any one of the possible p_rth roots. Nothing said so far tells which one to choose. As a result we shall wind up with a number of different values for $\alpha_1, \ldots, \alpha_{p_r}$, and generally more than n of these. In time-honored tradition we distinguish those which are actually roots from those which are not by substituting them in the original equation, discarding the ones that do not satisfy it. In this way our procedure exhibits $\alpha_1, \ldots, \alpha_n$, but it might have produced a few strays along the way and these had to be eliminated by this trial and error. However, the procedure will assure us that all α's which are roots will be in our set and no root will be left out. This situation arises in the solution of the general cubic (see p. 176). As pointed out in Chapter I, the rather glib assurance that we may take any algebraic number in radicals, substitute it into $p(x)$, and see whether it is a root might itself turn into a not-quite-trivial problem, but luckily it is solvable. (Remember, for example, that $\sqrt[3]{1 + \frac{2}{3}\sqrt{\frac{7}{3}}} + \sqrt[3]{1 - \frac{2}{3}\sqrt{\frac{7}{3}}} = 1$.)

Now for the answer to question (1): We describe a process that tells us how to evaluate $\varphi_{r-1}(\zeta_r)$ in terms of the functions $\varphi_{r-2}(\zeta_{r-1})$; then the $\varphi_{r-2}(\zeta_{r-1})$ in terms of the $\varphi_{r-3}(\zeta_{r-2})$, and so on, as follows.

Again we shall work on the general quartic before describing the procedure in general. Here we have $r - 1 = 3$, $p_{r-1} = 2$, so $\zeta_3 = +1$ or -1 and $H_3 = G_2/G_3 = \mathfrak{B}/\mathfrak{H} = \{[(1)], [(13)(24)]\}$, where $[(1)] = \{(1), (12)(34)\}$ and $[(13)(24)] = \{(13)(24), (14)(23)\}$.

If now $\theta \in [1]$, then $\theta\varphi_3(\zeta_4) = \varphi_3(\zeta_4)$. For example, $(12)(34)\varphi_3(-1) = (12)(34)(\alpha_1 - \alpha_2)^2 = (\alpha_2 - \alpha_1)^2 = \varphi_3(-1)$. However, if $\theta \in [(13)(24)]$, we get new functions that are conjugate to $\varphi_3(\zeta_4)$. For example, $(13)(24)\varphi_3(-1) = (13)(24)(\alpha_1 - \alpha_2)^2 = $ _____, and $(14)(23)\varphi_3(-1) = $ _____ $= $ _____ $= (13)(24)\varphi_3(-1)$, so we get the same new function for each element of the coset $[13)(24)]$.

Let

$$\psi_{3,1} = \varphi_3(\zeta_4),$$

$$\psi_{3,2} = (13)(24)\varphi_3(\zeta_4),$$

or, in alternative notation, $\{\psi_{3,1}, \psi_{3,2}\} = \mathfrak{B}\{\varphi_3(\zeta_4)\}$. (Of course, $\psi_{3,1}$ and $\psi_{3,2}$ are still functions of the α's and ζ_4, but the arguments will not be indicated at each step.) We already know that each of these new functions is left unchanged by any element of G_3 (namely $\mathfrak{H}$) and is therefore an element of F_{G_3}. Next we let

$$f_{3,1}(x) = \psi_{3,1} + x\psi_{3,2},$$

$$f_{3,2}(x) = \psi_{3,2} + x\psi_{3,1} = (13)(24)f_{3,1}(x)$$

[the permutation $(13)(24)$ is applied to the second subscript of the ψ's]. So

$$f_{3,1}(1) = \psi_{3,1} + \psi_{3,2} = \varphi_3(\zeta_4) + (13)(24)\varphi_3(\zeta_4)$$

$$= (\alpha_1 + \zeta_4\alpha_2)^2 + (\alpha_3 + \zeta_4\alpha_4)^2,$$

$$f_{3,1}(-1) = \psi_{3,1} - \psi_{3,2},$$

$$f_{3,2}(-1) = \psi_{3,2} - \psi_{3,1},$$

4.5 HOW TO SOLVE A SOLVABLE EQUATION

and each $f_{3,j}$ is invariant under G_3. The new functions

$$\varphi_2(\zeta_3) = (f_{3,1}(\zeta_3))^2 = (f_{3,2}(\zeta_3))^2 \tag{1}$$

should then, by the general results on cyclic equations, be invariant under $\mathfrak{B}$. To check this, we substitute $\zeta_3 = +1$ or -1 and see that

$$\varphi_2(1) = (\psi_{3,1} + \psi_{3,2})^2$$

$$= (\varphi_3(\varphi_4) + (13)(24)\varphi_3(\zeta_4))^2$$

$$= ((\alpha_1 + \zeta_4\alpha_2)^2 + (\alpha_3 + \zeta_4\alpha_4)^2)^2,$$

$$\varphi_2(-1) = ((\alpha_1 + \zeta_4\alpha_2)^2 - (\alpha_3 + \zeta_4\alpha_4)^2)^2,$$

where ζ_4 may be $+1$ or -1. It is easily seen on inspection that each of these new functions is invariant under $\mathfrak{B}$, which is the G_2 of this example, and its value is therefore an element of F_{G_2}. [In fact, taking $\zeta_4 = +1$ shows that in this example $\varphi_3(1)$ is symmetric in the roots and so even an element of the ground field F, but this is not true in general.] Note also that each φ_2 depends both on ζ_3 and on ζ_4. Equation (1) shows that $f_{3,1}(\zeta_3)$ and $f_{3,2}(\zeta_3)$ are obtained from $\varphi_2(\zeta_3)$ by a square root, and the ψ's are obtained from these as before:

$$\sum_{\zeta_r} f_{3,1}(\zeta_3) = (\psi_{3,1} + \psi_{3,2}) + (\psi_{3,1} - \psi_{3,2}) = 2\psi_{3,1},$$

$$\sum_{\zeta_3} f_{3,2}(\zeta_3) = (\psi_{3,2} + \psi_{3,1}) + (\psi_{3,2} - \psi_{3,1}) = 2\psi_{3,2}.$$

So if the values of the φ_2 were known, then the $f_{3,j}(\zeta_3)$, the $\psi_{3,j}$, the φ_3 (by p. 162), the $f_{4,j}$ (by p. 163), and finally the α's could be calculated.

Now for the next step:

We have $r - 2 = 2$, $p_{r-2} = p_2 = 3$, so $\zeta_2 = 1$ or ω or ω^2 (where $\omega = -\frac{1}{2} + \frac{1}{2}\sqrt{-3}$) and

$$H_2 = G_1/G_2 = \mathfrak{A}/\mathfrak{B} = \{[(1)], [(123)], [(132)]\}.$$

Each coset in H_2 contains four elements and there are three cosets.

Let

$$\psi_{2,1} = \varphi_2(\zeta_3),$$

$$\psi_{2,2} = (123)\varphi_2(\zeta_3),$$

$$\psi_{2,3} = (132)\varphi_2(\zeta_3)$$

and

$$f_{2,1}(x) = \psi_{2,1} + x\psi_{2,2} + x^2\psi_{2,3},$$

$$f_{2,2}(x) = \psi_{2,2} + x\psi_{2,3} + x^2\psi_{2,1} = (123)f_{2,1}(x),$$

$$f_{2,3}(x) = \psi_{2,3} + x\psi_{2,1} + x^2\psi_{2,2} = (123)^2 f_{2,1}(x),$$

$$= (132)f_{2,1}(x).$$

If we remember that $1 + \omega + \omega^2 = 0$ and $\omega^3 = 1$, then the last set of equations shows that

$$\sum_{\zeta_2} f_{2,1}(\zeta_2) = 3\psi_{2,1}, \qquad \psi_{2,1} = \tfrac{1}{3}\sum_{\zeta_2} f_{2,1}(\zeta_2),$$

$$\sum_{\zeta_2} f_{2,2}(\zeta_2) = 3\psi_{2,2}, \qquad \psi_{2,2} = \tfrac{1}{3}\sum_{\zeta_2} f_{2,2}(\zeta_2),$$

$$\sum_{\zeta_2} f_{2,3}(\zeta_2) = 3\psi_{2,3}, \qquad \psi_{2,3} = \tfrac{1}{3}\sum_{\zeta_2} f_{2,3}(\zeta_2).$$

Again raising $f_{2,j}(\zeta)$ to the p_2th power, we get

$$(f_{2,1}(\zeta_2))^3 = (f_{2,2}(\zeta_2))^3 = (f_{2,3}(\zeta_2))^3,$$

so that $(f_{2,1}(\zeta_2))^3$ is invariant under G_1 and must be an element of F_{G_1}, and the $f_{2,j}(\zeta_2)$ are obtained from this by extraction of a cube root.

A next and last repetition of this procedure would lead us to a function invariant under G_0 and therefore in the ground field F. In this example, G_0 is the symmetric

4.5 HOW TO SOLVE A SOLVABLE EQUATION

group $\mathfrak{S}_4$, so the last functions we obtain will be symmetric in the α's and thus can also be evaluated by time-tested, although somewhat laborious, procedures. Then working backward from these values as indicated on pages 163, 165, and 167, we shall eventually find the roots $\alpha_1, \ldots, \alpha_n$. We repeat that this process, although effective, introduces extraneous answers that must be eliminated by careful checking with the original polynomial $p(x)$.

We now return to the general case. Suppose the $\varphi_{i-1}(\zeta_i)$ are constructed as formal expressions in $\alpha_1, \ldots, \alpha_n$ and roots of unity and that we have

$$\cdots G_{i-1} \underset{p_i}{\triangleright} G_i \cdots .$$

$H_i = G_{i-1}/G_i = \{[(1)], [\theta], [\theta^2], \ldots, [\theta^{p_i-1}]\}$, where θ is some properly chosen permutation in G_{i-1}. Since H_i is cyclic of prime order p_i, we can be sure that G_{i-1} will contain a suitable permutation θ. (Can there be several such θ's? _____ Can one take as θ any permutation of G_{i-1} which is not also an element of G_i? _____ _____)

Let

$$\psi_{i,1} = \varphi_i(\zeta_{i+1}),$$

$$\psi_{i,2} = \theta\varphi_i(\zeta_{i+1}) = \theta\psi_{i,1},$$

$$\psi_{i,3} = \theta^2\psi_{i,1},$$

$$\vdots$$

$$\psi_{i,p_i} = \theta^{p_i-1}\psi_{i,1}, \qquad i = 1, \ldots, r-1,$$

and

$$f_{i,1}(x) = \psi_{i,1} + x\psi_{i,2} + x^2\psi_{i,3} + \cdots + x^{p_i-1}\psi_{i,p_i},$$

$$f_{i,2}(x) = \theta f_{i,1}(x),$$

$$\vdots$$

$$f_{i,j}(x) = \theta^j f_{i,1}(x), \qquad j = 1, \ldots, p_i, \quad i = 1, \ldots, r-1.$$

We already know that each φ_i is invariant under the permutations of G_i, so since G_i is normal in G_{i-1} and its cosets are generated by θ we can be sure that each $f_{i,j}(x)$ is invariant under G_i. Moreover, since the period of θ is a prime, all $f_{i,j}(x)$, and also all $f_{i,j}(\zeta_i)$ are different. As in the examples, we have

$$\sum_{\zeta_i} f_{i,j}(\zeta_i) = p_j \psi_{i,j'}, \qquad \psi_{i,j} = \underline{\qquad\qquad},$$

so the $\psi_{i,j}$ can be recovered from the $f_{i,j}$.

Also, again since $\theta^{p_i} = (1)$, if we let

$$\varphi_{i-1}(\zeta_i) = (f_{i,1}(\zeta_i))^{p_i} = \cdots = (f_{i,j}(\zeta_i))^{p_i} = \cdots = (f_{i,p_i}(\zeta_i))^{p_i},$$

then $\varphi_{i-1}(\zeta_i)$ is invariant under G_{i-1}, making $\varphi_{i-1}(\zeta_i)$ an element of $F_{G_{i-1}}$ from which $f_{i,j}(\zeta_i)$ is obtained by a p_ith root.

This process continues until we reach G_0, the group of the equation, and F_{G_0}, which is of course the field of coefficients F. If now G_0 is the symmetric group $\mathfrak{S}_n$, then the functions $\varphi_0(\zeta_1)$ will be symmetric in $\alpha_1, \ldots, \alpha_n$ and so can be evaluated by standard methods [W, p. 107]. If G_0 is not symmetric, a detour is necessary. First we shall give a description of these extra steps, then carry out the entire process described above on the general cubic equation, whose group is the symmetric group $\mathfrak{S}_3$ and on a particular cubic equation whose group is $\mathfrak{A}_3$. So suppose φ_1 is any function of $\alpha_1, \ldots, \alpha_n$ (and possibly various roots of unity) which is invariant under the group G_0. We know that $G_0 < \mathfrak{S}_n$, so let us form the k distinct functions $\varphi = \varphi_1, \varphi_2, \ldots, \varphi_k$, which can be obtained from φ by applying the $n!$ different permutations of $\mathfrak{S}_n$ to the subscripts of the α's and discarding duplications. (For instance, if $\varphi = (\alpha_1 - \alpha_2)^2$, then $(12)\varphi = (\alpha_2 - \alpha_1)^2 = \varphi$, so $(12)\varphi$ would be discarded.) Even though G_0 is not necessarily a normal subgroup of $\mathfrak{S}_n$, we would find that $k = \mathrm{ind}_{\mathfrak{S}_n} G_0$, but it is not necessary now to know this. Next we form the polynomial

$$q(y) = (y - \varphi_1)(y - \varphi_2) \cdots (y - \varphi_k)$$

$$= b_0 + b_1 y + \cdots + b_k y^k,$$

4.5 HOW TO SOLVE A SOLVABLE EQUATION

where each b_i is symmetric in $\varphi_1, \ldots, \varphi_k$. Now if we compare the functions $\varphi_1, \ldots, \varphi_k$ with $\sigma\varphi_1, \ldots, \sigma\varphi_k$, where $\sigma \in \mathfrak{S}_n$, we see that the second set is just a permutation of the first: For, by construction, $\varphi_1, \ldots, \varphi_k$ are all functions obtainable from φ_1 by permutations of $\alpha_1, \ldots, \alpha_n$ and if $\sigma\varphi_i = \sigma\varphi_j = \varphi_k$, then $\varphi_i = \sigma^{-1}\varphi_k = \varphi_j$, so σ is a one-to-one mapping of $\varphi_1, \ldots, \varphi_k$ onto $\sigma\varphi_1, \ldots, \sigma\varphi_k$ — that is, a permutation of these functions. As a result it follows that any function that is symmetric in $\varphi_1, \ldots, \varphi_k$ is unchanged by any $\sigma \in \mathfrak{S}_n$, so it must be symmetric in $\alpha_1, \ldots, \alpha_n$. Hence each $b_0, \ldots, b_n$ is symmetric in $\alpha_1, \ldots, \alpha_n$ and therefore can be calculated. The polynomial $q(y) = b_0 + \cdots + b_k y^k$ is then known.

Our original function φ_1 is invariant under the group G_0 of the given polynomial $p(x)$ and is therefore an element of the ground field F. It is also a root of $q(y)$, so $q(y)$ must have at least one linear factor in $F_\varepsilon[y]$. Part of the hypothesis is that there should be a factorization algorithm for $F_\varepsilon[y]$. Using this, we factor

$$q(y) = (y - \beta_1) \cdots (y - \beta_m) r(y),$$

where $\beta_1, \ldots, \beta_m \in F_\varepsilon$ are the possible values of φ_1. Working backward from φ_1 to $\alpha_1, \ldots, \alpha_n$ we eventually find all the roots of the given equation. Trial and error will eliminate redundant roots.

It is perhaps worth remarking that the hypothesis that we may add various nth roots of unity to F is avoidable if F is large enough. We could have defined

$$f_{i,1} = m_1 \psi_{i,1} + m_2 \psi_{i,2} + \cdots + m_{p_i} \psi_{i,p_i}$$

and picked $m_1, \ldots, m_{p_i}$ so as to be sure that $f_{i,1}, \ldots, f_{i,p_i}$ are all different and that we can recover $\psi_{i,1}, \ldots, \psi_{i,p_i}$ by solving. This is just what we shall do in the construction of a resolvent of $p(x)$. In any actual example, however, this would complicate the solution of $p(x)$ even more than the introduction of the roots of unity, so it seemed best not to mention it when first explaining the procedure. On the other hand, it shows that the hypothesis that there be a factorization algorithm in $F_\varepsilon[x]$ is superfluous, for F_ε is not needed in this case. We only need a factorization algorithm for $F[x]$.

(Remember that primitive roots of unity cannot always be added freely: If $F = \{0, 1\}$ is the field of two elements with characteristic 0, the assumption that there is a primitive square root of unity quickly leads to a contradiction. For suppose we form $F(a)$, where $a \neq 1$, $a^2 = 1$, and a is a new element added to F. Then $(a + 1)^2 = a^2 + 2a + 1 = a^2 + 1 = 1 + 1 = 0$, so $a + 1 = 0$ and $a = 1$, contradicting the hypothesis that $a \neq 1$.)

Notice also that if G_i/G_{i+1} is not cyclic, then we cannot be sure that $(f_{i,j}(\zeta_{p_i}))^{p_i}$ is an element of $F_{G_{i-1}}$, so this method of solution breaks down. The assumption that the group of the given polynomial be solvable is therefore essential to this method of solving equations. Without the results of Section 4.3 we could, however, not conclude that equations with unsolvable groups are themselves unsolvable, for we might simply not have come up with a sufficiently ingenious method. Lagrange realized this in the late eighteenth century, but it was not until the nineteenth century when Galois theory was developed that the work of Galois and Abel showed that the general equations of degree 5 and higher are unsolvable in radicals.

We close the section with some examples. The main purpose of these examples is to illustrate the general procedure. Anyone interested only in the general solution of the cubic need not get involved in Galois theory. The systematic solution requires much detailed work, so please be patient!

EXAMPLE (1). The general cubic over Q.

Let

$$p(x) = c_0 + c_1 x + c_2 x^2 + x^3,$$

$$F = Q,$$

$$F_\varepsilon = Q(\omega).$$

We have

$$G_0 = \mathfrak{S}_3, \qquad G_1 = \mathfrak{A}_3, \qquad G_2 = (1),$$

4.5 HOW TO SOLVE A SOLVABLE EQUATION

so $r = 2$, and

$$\mathfrak{S}_3 \underset{p_1=2}{\triangleright} \mathfrak{A}_3 \underset{p_2=3}{\triangleright} (1),$$

$$H_2 = \mathfrak{A}_3 = \{(1), (123), (132)\},$$

$$H_1 = \mathfrak{S}_3/\mathfrak{A} = \{[(1)], [(12)]\} = \{\{(1), (123), (132)\}, \{(12), (13), (23)\}\},$$

$$\zeta_2 = 1 \text{ or } \omega \text{ or } \omega^2, \qquad \zeta_1 = 1 \text{ or } -1,$$

$$f_{2,1}(x) = \alpha_1 + x\alpha_2 + x^2\alpha_3,$$

$$f_{2,2}(x) = \alpha_2 + x\alpha_3 + x^2\alpha_1 = (123)f_{2,1}(x),$$

$$f_{2,3}(x) = \alpha_3 + x\alpha_1 + x^2\alpha_2 = (132)f_{2,1}(x),$$

$$f_{2,1}(1) = \alpha_1 + \alpha_2 + \alpha_3 = -c_2$$

$$= f_{2,2}(1) = f_{2,3}(1),$$

$$\varphi_1(1) = -c_2^3,$$

$$f_{2,1}(\omega) = \alpha_1 + \alpha_2 + \omega^2\alpha_3,$$

$$f_{2,2}(\omega) = \underline{\hspace{4cm}},$$

$$f_{2,3}(\omega) = \underline{\hspace{4cm}},$$

$$\varphi_1(\omega) = (f_{2,1}(\omega))^3 = (f_{2,2}(\omega))^3 = (f_{2,3}(\omega))^3$$

$$= \underline{\hspace{6cm}}$$

$$= \alpha_1^3 + \alpha_2^3 + \alpha_3^3 + 3\omega(\alpha_1\alpha_3^2 + \alpha_3\alpha_2^2 + \alpha_2\alpha_1^2) + 3\omega^2(\alpha_1\alpha_2^2 + \alpha_2\alpha_3^2 + \alpha_3\alpha_1^2)$$

$$+ 6\alpha_1\alpha_2\alpha_3,$$

$$f_{2,1}(\omega^2) = \alpha_1 + \omega^2\alpha_2 + \omega^4\alpha_3 = \alpha_1 + \omega^2\alpha_2 + \omega\alpha_3,$$

$$f_{2,2}(\omega^2) = \underline{\hspace{5cm}},$$

$$f_{2,3}(\omega^2) = \underline{\hspace{5cm}},$$

$$\varphi_1(\omega^2) = (f_{2,1}(\omega^2))^3 = (f_{2,2}(\omega^2))^3 = (f_{2,3}(\omega^2))^3$$

$$= \underline{\hspace{7cm}}$$

$$= \alpha_1^3 + \alpha_2^3 + \alpha_3^3 + 3\omega(\alpha_1\alpha_2^2 + \alpha_2\alpha_3^2 + \alpha_3\alpha_1^2) + 3\omega^2(\alpha_1\alpha_3^2 + \alpha_3\alpha_2^2 + \alpha_2\alpha_1^2)$$

$$+ 6\alpha_1\alpha_2\alpha_3.$$

On comparing $\varphi_1(\omega)$ and $\varphi_1(\omega^2)$, notice that $\varphi_1(\omega) = (12)\varphi_1(\omega^2)$. Continuing, we have

$$\psi_{1,1}(1) = \varphi_1(1) = -c_2^3,$$

$$\psi_{1,1}(\omega) = \varphi_1(\omega),$$

$$\psi_{1,2}(\omega) = (12)\varphi_1(\omega) = \varphi_1(\omega^2),$$

$$\psi_{1,1}(\omega^2) = \varphi_1(\omega^2),$$

$$\psi_{1,2}(\omega^2) = (12)\varphi_1(\omega^2) = (12)(12)\varphi_1(\omega)$$

$$= \varphi_1(\omega),$$

$$f_{1,1}(1, \omega) = \psi_{1,1}(\omega) + \psi_{1,2}(\omega)$$

$$= \varphi_1(\omega) + \varphi_1(\omega^2)$$

$$= 2(\alpha_1^3 + \alpha_2^3 + \alpha_3^3) + 3\omega(\underline{\hspace{3cm}})$$

$$+ 3\omega^2(\underline{\hspace{3cm}})$$

$$+ 12\alpha_1\alpha_2\alpha_3$$

$$= 2(\alpha_1^3 + \alpha_2^3 + \alpha_3^3) - 3(\underline{\hspace{3cm}})$$

$$+ 12\alpha_1\alpha_2\alpha_3.$$

4.5   HOW TO SOLVE A SOLVABLE EQUATION

So $f_{1,1}(1, \omega)$ is symmetric and can be found in terms of c_0, c_1, c_2. Calculation of this symmetric function gives

$$f_{1,1}(1, \omega) = 2(-3c_0 + 3c_1c_2 - c_2^3) - 3(3c_0 - c_1c_2) - 12c_0$$

$$= -27c_0 + 9c_1c_2 - 2c_2^3.$$

Continuing the general procedure we have

$$f_{1,1}(1, \omega^2) = \psi_{1,1}(\omega^2) + \psi_{1,2}(\omega^2) = \varphi_1(\omega^2) + \varphi_1(\omega)$$

$$= f_{1,1}(1, \omega),$$

$$f_{1,1}(-1, \omega) = \psi_{1,1}(\omega) - \psi_{1,2}(\omega)$$

$$= \varphi_1(\omega) - \varphi_1(\omega^2)$$

$$= 3\omega(\underline{\hspace{3cm}})$$

$$+3\omega^2(\underline{\hspace{3cm}})$$

$$= 3\sqrt{-3}(A - B),$$

where

$$A = \alpha_1\alpha_3^2 + \alpha_3\alpha_2^2 + \alpha_2\alpha_1^2, \qquad B = \alpha_1\alpha_2^2 + \alpha_2\alpha_3^2 + \alpha_3\alpha_1^2,$$

and we are using the facts that $\omega^2 = -\omega - 1$ and $2\omega + 1 = \sqrt{-3}$. Also,

$$f_{1,2}(-1, \omega^2) = \psi_{1,2}(\omega^2) - \psi_{1,1}(\omega^2)$$

$$= \varphi_1(\omega) - \varphi_1(\omega^2)$$

$$= f_{1,1}(-1, \omega),$$

$$\psi_0(-1) = (f_{1,1}(-1, \omega))^2 = 27(A - B)^2$$

$$= -27(\alpha_1\alpha_3^2 + \alpha_3\alpha_2^2 + \alpha_2\alpha_1^2 - \alpha_1\alpha_2^2 - \alpha_2\alpha_3^2 - \alpha_3\alpha_1^2)^2$$

$$= -27[(\alpha_1 - \alpha_2)(\alpha_2 - \alpha_3)(\alpha_1 - \alpha_3)]^2$$

$$= -27\Delta,$$

where Δ as usual denotes the discriminant of the cubic and is symmetric in $\alpha_1, \alpha_2, \alpha_3$. The discriminant therefore appears here in a very natural manner, not as an artificially contrived function of the roots which just "happens" to work. Calculation of this symmetric function yields

$$\Delta = c_2^2 c_1^2 + 18c_2 c_1 c_0 - 4c_1^3 - 4c_2^3 c_0 - 27c_0^2,$$

so that

$$3\sqrt{-3(A-B)} = \pm\sqrt{-27\Delta} = f_{1,1}(-1, \omega),$$

$$\sum_{\zeta_1} f_{1,1}(\zeta_1, \omega) = f_{1,1}(1, \omega) + f_{1,1}(-1, \omega) = 2\psi_{1,1}(\omega)$$

$$= -27c_0 + 9c_1 c_2 - 2c_2^3 \pm \sqrt{-27\Delta}.$$

Hence

$$f_{2,1}(\omega) = \sqrt[3]{\psi_{1,1}(\omega)} = \sqrt[3]{c} \text{ or } \omega\sqrt[3]{c} \text{ or } \omega^2\sqrt[3]{c},$$

where $c^{\pm} = \frac{1}{2}(-27c_0 + 9c_1 c_2 - 2c_2^3 \pm \sqrt{-27\Delta})$.

Similarly, we find

$$f_{2,1}(\omega^2) = \sqrt[3]{c} \text{ or } \omega\sqrt[3]{c} \text{ or } \omega^2\sqrt[3]{c},$$

so that

$$3\alpha_1 = \sum_{\zeta_2} f_{2,1}(\zeta_2) = f_{2,1}(1) + f_{2,1}(\omega) + f_{2,1}(\omega^2)$$

$$= -c_2 + \begin{cases} \sqrt[3]{c^{\pm}} \\ \text{or} \\ \omega\sqrt[3]{c^{\pm}} \\ \text{or} \\ \omega^2\sqrt[3]{c^{\pm}} \end{cases} + \begin{cases} \sqrt[3]{c^{\pm}} \\ \text{or} \\ \omega\sqrt[3]{c^{\pm}} \\ \text{or} \\ \omega^2\sqrt[3]{c^{\pm}}, \end{cases}$$

4.5 HOW TO SOLVE A SOLVABLE EQUATION

$$3\alpha_2 = \sum_{\zeta_2} f_{2,2}(\zeta_2) = f_{2,2}(\omega) + f_{2,2}(\omega^2)$$

$$= -c_2 + \left\{ \frac{\underline{\qquad\qquad}}{\underline{\qquad\qquad}} \right\} + \left\{ \frac{\underline{\qquad\qquad}}{\underline{\qquad\qquad}} \right\},$$

$$3\alpha_3 = \sum_{\zeta_2} f_{2,3}(\zeta_3) = -c_2 + \left\{ \frac{\underline{\qquad\qquad}}{\underline{\qquad\qquad}} \right\} + \left\{ \frac{\underline{\qquad\qquad}}{\underline{\qquad\qquad}} \right\}.$$

This solution for α_1, α_2, α_3 of course must contain extraneous roots, since $c^{\pm}$ in general takes on two different values, depending on the sign of Δ, and which of the cube roots of $c^{\pm}$ is then chosen. The extraneous answers are eliminated by trail and error. For example, we must have

$$3\alpha_1 + 3\alpha_2 + 3\alpha_3 = -3c_2.$$

It is advisable to check that taking

$$3\alpha_1 = -c_2 + \sqrt[3]{c^+} + \sqrt[3]{c^-},$$

$$3\alpha_2 = -c_2 + \omega\sqrt[3]{c^+} + \omega^2\sqrt[3]{c^-},$$

$$3\alpha_3 = -c_2 + \omega^2\sqrt[3]{c^+} + \omega\sqrt[3]{c^-},$$

does yield a satisfactory solution. Having thus listed three roots of the equation, we know that there cannot be any more and it is unnecessary to test the remaining possibilities. The resulting formulas are known as *Cardan's solution* of the cubic.

EXAMPLE (2). Solve the equation $p(x) = 0$ over Q, where

$$p(x) = x^3 - 3x + 1,$$

whose group is the cyclic group $\mathfrak{A}_3 = \{(1), (123), (132)\}$. Here we have $G = G_0 = \mathfrak{A}_3 \underset{p_1 = 3}{\triangleright} G_1 = \{(1)\}$, so

$$r = 1,$$

$$H_1 = \mathfrak{A}_3/(1) = \mathfrak{A}_3,$$

$$\zeta_1 = 1, \omega, \omega^2,$$

$$f_{1,1}(x) = \alpha_1 + x\alpha_2 + x^2\alpha_3,$$

$$f_{1,2}(x) = \text{_____},$$

$$f_{1,3}(x) = \text{_____},$$

$$c_0 = +1, \quad c_1 = -3, \quad c_2 = 0.$$

In this example G is not the full symmetric group, so $\varphi_0(\zeta_1)$ will not be symmetric in $\alpha_1, \alpha_2, \alpha_3$.

We get for $i = 1, 2, 3$ that

$$\varphi_0(1) = (f_{1,i}(1))^3 = (\alpha_1 + \alpha_2 + \alpha_3)^3 = 0,$$

$$\varphi_0(\omega) = (f_{1,i}(\omega))^3 = \alpha_1^3 + \alpha_2^3 + \alpha_3^3 + 3\omega B + 3\omega^2 A + 6\alpha_1\alpha_2\alpha_3,$$

$$\varphi_0(\omega^2) = (f_{1,i}(\omega^2))^3 = \alpha_1^3 + \alpha_2^3 + \alpha_3^3 + 3\omega A + 3\omega^2 B + 6\alpha_1\alpha_2\alpha_3.$$

The group of $p(x)$ is cyclic, so we must have $\varphi_0(\omega) \in Q(\omega)$ and $\varphi_0(\omega^2) \in Q(\omega)$. To evaluate them, we first note that

$$\alpha_1^3 + \alpha_2^3 + \alpha_3^3 = -3c_0 + 3c_1c_2 - c_2^3 = -3 + 0 = -3,$$

$$\sum \alpha_1\alpha_2^2 = -3c_0 + c_1c_2 = 3,$$

and

$$\alpha_1\alpha_2\alpha_3 = -c_0 = -1.$$

4.5 HOW TO SOLVE A SOLVABLE EQUATION

So

$$\varphi_0(\omega) = -3 + 3\omega B + 3\omega^2 A - 6 = -9 + 3\omega B + 3\omega^2 A,$$

$$\varphi_0(\omega^2) = -9 + 3\omega A + 3\omega^2 B.$$

Let us first work with $\varphi_0(\omega)$. Under the symmetric group $\mathfrak{S}_3$ this has only one conjugate, for we find that

$$\varphi_0(\omega) = (123)\varphi_0(\omega) = (132)\varphi_0(\omega),$$

$$(12)\varphi_0(\omega) = (13)\varphi_0(\omega) = (23)\varphi_0(\omega) = \varphi_0(\omega^2),$$

so the equation

$$(y - \varphi_0(\omega))(y - \varphi_0(\omega^2)) = 0$$

must have roots in $Q(\omega)$, if the group of the given equation is in fact $\mathfrak{A}_3$. Expanding yields

$$y^2 = (\varphi_0(\omega) + (12)\varphi_0(\omega))y + \varphi_0(\omega) \cdot (12)\varphi_0(\omega)$$

$$= (2\sum \alpha_i^3 + 3(\omega + \omega^2)(A + B) + 12\alpha_1\alpha_2\alpha_3)$$

$$+ (-9 + 3\omega B + 3\omega^2 A)(-9 + 3\omega A + 3\omega^2 B)$$

$$= y^2 - (2) \cdot (-3) - 3(3) + 12(-1)$$

$$+ 9(9 - 3\omega B - 3\omega^2 A - 3\omega A + \omega^2 AB + A^2 - 3\omega^2 B + B^2 + \omega AB)$$

$$= y^2 - 27y + 9(9 - 3(\omega + \omega^2)B - 3(\omega + \omega^2)A + (\omega + \omega^2)AB + A^2 + B^2)$$

$$= y^2 - 27y + 9(9 + 3(A + B) + A^2 - AB + B^2)$$

$$= y^2 - 27y + 9(9 + 3(A + B) + (A + B)^2 - 3AB). \tag{2}$$

Using

$$A + B = \sum \alpha_1 \alpha_2^2 = 3c_0 - c_1 c_2 = 3,$$

$$(A + B)^2 = 9,$$

$$(A - B)^2 = \Delta = c_2^2 c_1^2 + 18c_2 c_1 c_0 - 4c_1^3 - 4c_2^3 c_0 - 27c_0^3$$

$$= -4(-3)^3 - 27 = 81,$$

$$4AB = (A + B)^2 - (A - B)^2 = 9 - 81 = -72,$$

$$AB = -18,$$

we see that the right side of (2) equals

$$y^2 - 27y + 9(9 + 3(3) + 9 - 3(-18))$$

$$= y^2 - 27y + 9(81)$$

$$= y^2 - 27y + 27^2 = 0, \tag{2}$$

so that $y = 27\omega$ or $27\omega^2$. If we take $\varphi_0(\omega) = 27\omega$ and $\varphi_0(\omega^2) = 27\omega^2$, then

$$\left. \begin{array}{l} f_{1,i}(\omega) = \sqrt[3]{27\omega} \\[2mm] f_{1,i}(\omega^2) = \sqrt[3]{27\omega^2} \end{array} \right\} \text{(different cube roots for different values of } i = 1, 2, 3)$$

and

$$3\alpha_i = f_{1,i}(1) + f_{1,i}(\omega) + f_{1,i}(\omega^2) = 0 + \sqrt[3]{27\omega} + \sqrt[3]{27\omega^2}$$

so

$$\alpha_i = (\sqrt[3]{\omega} + \sqrt[3]{\omega^2}).$$

More explicitly, we can list the roots as

$$\alpha_1 = (\sqrt[3]{\omega} + \omega^2\sqrt[3]{\omega^2}) \quad = (\omega^{1/3} + \omega^{-1/3}),$$

$$\alpha_2 = (\omega\sqrt[3]{\omega} + \omega\sqrt[3]{\omega^2}) \quad = (\omega^{4/3} + \omega^{-4/3}),$$

$$\alpha_3 = (\omega^2\sqrt[3]{\omega} + \sqrt[3]{\omega^2}) \quad = (\omega^{2/3} + \omega^{-2/3}).$$

In this last set we had to choose the cube roots carefully to get the correct roots. For example, we could not have chosen them as $\alpha = \omega^{1/3} + \omega^{-2/3}$, because this would not have satisfied the equation. Incidentally, notice that $\alpha_1, \alpha_2, \alpha_3$ are all real.

4.6   RULER-AND-COMPASS CONSTRUCTIONS

In Section 4.3 we showed that certain equations cannot be solved in radicals by characterizing the class of all solvable equations and showing that there are some equations which do not belong to this class. In the present section we shall in turn characterize the class of all those equations whose roots are constructible by ruler and compass and shall see immediately that there are many equations which do not belong to this class. For example, the equation $x^3 - 2 = 0$ does not have constructible roots. Therefore, we cannot construct a length $\alpha = \sqrt[3]{2}$ by ruler and compass alone, which shows that the classical problem of the duplication of a cube is unsolvable.

Galois theory is not essential to the proof that certain constructions are impossible, but it does add additional insight [W, p. 154; C-R, p. 127]. It is necessary to give a precise definition of "ruler-and-compass" construction. The notion goes back to the ancient Greeks and these constructions turn out to be particularly suited to algebraic treatment. (Some other methods and machines for constructions are known [C-R, pp. 138, 146].)

The allowed tools and basic constructions are as follows.

Tools

(1) Paper and pencil.

(2) A straightedge: a ruler or other tool used to draw straight lines but with no divisions marked on it.

(3) A compass used to draw circles.

(4) A segment of unit length 1 marked on the paper.

Fundamental Constructions

(1) Given two points, we may draw a line of indefinite length through these two points.

(2) Given two points, we may draw the line segment joining them.

(3) Given a point P and a line segment AB, we may draw a circle with radius AB and center P.

Basic Assumption.   A point is determined by one intersection of two lines, a line and a circle, or two circles; a segment is determined by two points.

DEFINITION.   A *ruler-and-compass construction* is a finite sequence of the basic constructions listed above.

Among the constructions possible by ruler and compass are the following:

(1) To draw the perpendicular to a given line at a given point.

(2) Given a line L and a point P not on L, to draw a line through P and parallel to L.

(3) Draw segments of length n units. (Remember that we are given the length 1 unit.)

(4) To draw angles of 45° and 60°.

(5) Given segments of length a and b, to draw segments of lengths $a + b$ and $(a - b)$.

(6) Given segments of lengths a and b, to draw segments of lengths ab and a/b.

Procedure: See Figures 4.1 and 4.2 and describe the steps in the construction in the space next to each figure.

4.6 RULER-AND-COMPASS CONSTRUCTIONS

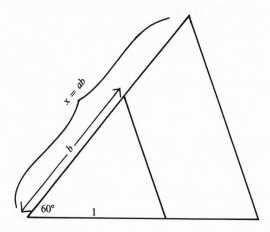

FIGURE 4.1

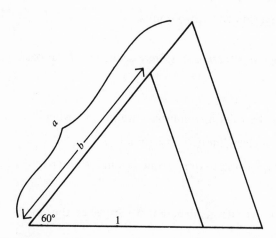

FIGURE 4.2

(7) Given a segment of length a, to draw a segment of length $\sqrt{a}$.

Procedure: See Figure 4.3.

    Proof that $x = \sqrt{a}$: _____

_____. ‖

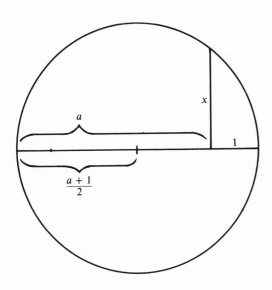

FIGURE 4.3

Notice that it is occasionally more convenient to use the Pythagorean theorem to construct certain lengths. For instance, this is true when we wish to construct $\sqrt{5}$ or $\sqrt{17}$, because $5 = 1^2 + 2^2$ and $17 = 1^2 + 4^2$.

For the algebraic characterization of ruler-and-compass construction it is convenient to introduce the following terminology:

DEFINITION. The *square-root field* $\mathscr{C}^{\sqrt{}}$ is the smallest field that contains Q and is closed under the square-root operation.

DEFINITION. The complex number $\alpha = a + ib$ is called *constructible* if we can construct a rectangle with sides of length a and b by a ruler-and-compass construction.

This means that we can locate the point α in the complex plane using ruler and compass only.

DEFINITION. A group G is called *constructible* if it has a composition series

$$G = G_0 \rhd G_1 \rhd G_2 \rhd \cdots \rhd G_{r-1} \rhd G_r = (1),$$

4.6 RULER-AND-COMPASS CONSTRUCTIONS

such that

$$|G_{i-1}/G_i| = 2, \qquad \text{for all } i, \qquad 1 \le i \le r.$$

The Jordan–Hölder theorem implies that in a constructible group *every* composition series is of this form.

We shall assume that the end points of the given segment of length 1 represent the complex numbers 0 and 1, respectively.

THEOREM. The complex number α is constructible $\Leftrightarrow \alpha \in \mathscr{C}^{\surd}$.

Proof. $\Rightarrow$ The proof is by induction on the number n of fundamental constructions used. If $n = 0$, then $\alpha = 0$ or 1 (one of the end points of the given segment) and $0, 1 \in \mathscr{C}^{\surd}$, so the theorem holds for $n = 0$. The induction hypothesis states that if γ is obtained by at most n fundamental constructions, then $\gamma \in \mathscr{C}^{\surd}$. A new point α can be constructed as (1) the intersection of two previously constructed lines, (2) a line and a circle, or (3) two circles. We assume that $\beta_1, \beta_2, \gamma_1, \gamma_2$ are obtained by at most n fundamental constructions and are therefore in $\mathscr{C}^{\surd}$.

Case 1. α is the intersection of two lines l_1 and l_2, where l_1 is obtained by joining two points $\beta_1, \gamma_1 \in \mathscr{C}^{\surd}$ and l_2 from $\beta_2, \gamma_2 \in \mathscr{C}^{\surd}$. The coordinates of $\alpha = a + ib$ are therefore $a = \underline{\hspace{5cm}}$ and $b = \underline{\hspace{3cm}}$ $\underline{\hspace{2cm}}$, so $\alpha \in \mathscr{C}^{\surd}$.

Case 2. α is an intersection of the line l through $\beta_1, \gamma_1 \in \mathscr{C}^{\surd}$ and the circle with center β_2 and passing through γ_2, with $\beta_2, \gamma_2 \in \mathscr{C}^{\surd}$. The coordinates of α are therefore $a = \underline{\hspace{6cm}}$ and $b = \underline{\hspace{3cm}}$ $\underline{\hspace{4cm}}$, so again $\alpha \in \mathscr{C}^{\surd}$.

Case 3. α is an intersection of the circles Γ_1 and Γ_2 with centers β_1, β_2 and passing through γ_1, γ_2, respectively. If $\beta_1 = b_1 + ib'_1$, $\beta_2 = b_2 + ib'_2$, $r_1 = |\beta_1 - \gamma_1|$, and $r_2 = |\beta_2 - \gamma_2|$, then we find α by solving simultaneously the equations

$$(x - b_1)^2 + (y - b'_1)^2 = r_1^2,$$

$$(x - b_2)^2 + (y - b'_2)^2 = r_2^2.$$

Subtracting the second from the first gives the linear equation

$$\underline{\hspace{7cm}}.$$

If we solve this simultaneously with the first equation, we need only the rational operations and $\sqrt{}$, so we obtain our solution $x, y \in \mathscr{C}^{\sqrt{}}$ and let $\alpha = x + iy$, so $\alpha \in \mathscr{C}^{\sqrt{}}$, These are the only possibilities for constructing a new point α, so the theorem holds.

$\Leftarrow$ Again we use induction, this time on the number n of operations $+, -, \times,$ $\div, \sqrt{}$, used to obtain α. The proof is left as an exercise using the constructions listed earlier. $\parallel$

THEOREM.   If α is a complex number and α is a root of $f(x)$ whose group is G_α, then

$$\alpha \in \mathscr{C}^{\sqrt{}} \Leftrightarrow G_\alpha \text{ is constructible.}$$

Proof. $\Rightarrow$ We again use induction on the number n of applications of $+, -, \times, \div,$ and $\sqrt{}$.

If $n = 0$, then $\alpha \in Q$, so $G_\alpha = (1)$ and is therefore constructible. Suppose the theorem holds for all β, γ constructed from rationals by at most n of the allowed operations, and that $\alpha = \beta \pm \gamma, \beta\gamma, \beta/\gamma$, or $\sqrt{\beta}$. Let Q_o be the splitting field of the polynomial $f_\alpha(x)$ and $Q_{\sqrt{\beta},\gamma}$ the splitting field of $f_\beta(x^2) \cdot f_\gamma(x)$. Then $\alpha \in Q_o$ and $\sqrt{\beta}, \gamma \in Q_{\sqrt{\beta},\gamma}$, so $\alpha \in Q_{\sqrt{\beta},\gamma}$ and

$$Q \lhd Q_\alpha \lhd Q_{\sqrt{\beta},\gamma}.$$

By the fundamental theorem of Galois theory, therefore,

$$G(Q_{\sqrt{\beta},\gamma}/Q) \rhd G(Q_{\sqrt{\beta},\gamma}/Q_\alpha) \rhd G(Q_{\sqrt{\beta},\gamma}/Q_{\sqrt{\beta},\gamma}) = (1),$$

and

$$G(Q_\alpha/Q) \cong G(Q_{\sqrt{\beta},\gamma}/Q)/G(Q_{\sqrt{\beta},\gamma}/Q_\alpha).$$

By the induction hypothesis G_β and G_γ are constructible. It follows that the splitting

4.6 RULER-AND-COMPASS CONSTRUCTIONS

fields $Q_\beta, Q_{\sqrt{\beta}}$, and Q_γ of $\beta, \sqrt{\beta}$, and γ are obtained from Q by sequences of quadratic extensions

$$Q \underset{2}{\lhd} Q(\beta_1) \underset{2}{\lhd} Q(\beta_2) \underset{2}{\lhd} \cdots \underset{2}{\lhd} Q(\beta_m) = Q_\beta \underset{2 \text{ or } 1}{\lhd} Q(\sqrt{\beta_1}) \underset{2 \text{ or } 1}{\lhd} \cdots \underset{2 \text{ or } 1}{\lhd} Q(\sqrt{\beta_m}) = Q_{\sqrt{\beta}},$$

$$Q \underset{2}{\lhd} Q(\gamma_1) \underset{2}{\lhd} \cdots \underset{2}{\lhd} Q(\gamma_n) = Q_\gamma.$$

These can be combined into one sequence,

$$Q \lhd Q(\beta_1) \cdots \lhd Q_{\sqrt{\beta}} \lhd Q_{\sqrt{\beta}}(\gamma_1) \lhd Q_{\sqrt{\beta}}(\gamma_2) \lhd \cdots \lhd Q_{\sqrt{\beta}}(\gamma_n) = Q_{\sqrt{\beta},\gamma},$$

where the degree of each field over the preceding field is 1 or 2. Therefore, the corresponding sequence of groups.

$$G(Q_{\sqrt{\beta},\gamma}/Q) \underset{2}{\rhd} G(Q_{\sqrt{\beta},\gamma}/Q(\beta_1)) \underset{2}{\rhd} \cdots \underset{2 \text{ or } 1}{\rhd} G(Q_{\sqrt{\beta},\gamma}/Q_\beta(\gamma_{n-1})) \underset{2 \text{ or } 1}{\rhd} \quad (1)$$

is a composition series for $G_{\sqrt{\beta},\gamma}$, the group of $Q_{\sqrt{\beta},\gamma}$ over Q, in which the order of each composition factor is 1 or 2, which shows that $G_{\sqrt{\beta},\gamma}$ is constructible. But $G(Q_\alpha/Q)$ is a homomorphic image of $G(Q_{\sqrt{\beta},\gamma}/Q)$ by (1), so $G(Q_\alpha/Q)$ is also constructible.

$\Leftarrow$ If G_2 is constructible, it has a composition series

$$G_\alpha \underset{2}{\rhd} G_1 \underset{2}{\rhd} \cdots \underset{2}{\rhd} G_l \underset{2}{\rhd} \cdots \rhd (1)$$

with corresponding fields

$$Q \underset{2}{\lhd} Q(\alpha_1) \underset{2}{\lhd} \cdots \underset{2}{\lhd} Q(\alpha_l) \underset{2}{\lhd} \cdots \lhd Q_\alpha,$$

where $Q(\alpha_i) = Q(\alpha)$. Since G_α is constructible, every element of each of these fields is obtained from the rationals by rational operations and $\sqrt{}$ only and is therefore in $\mathscr{C}^{\sqrt{}}$. In particular $\alpha \in \mathscr{C}^{\sqrt{}}$.   ‖

COROLLARY. α is constructible $\Leftrightarrow G(Q(\alpha)/Q)$ is constructible.

(Since $Q(\alpha)$ does not necessarily contain all the conjugates of α, it is not always a normal extension of Q, so $G(Q(\alpha)/Q)$ is not necessarily the same as the Galois group of the irreducible polynomial which has α as a root. For example, if $\alpha = \sqrt{1 + \sqrt{5}}$, then $\alpha' = \sqrt{1 - \sqrt{5}}$ is conjugate to α over Q, but $\alpha' \notin Q(\alpha)$, since α is real and α' is not.)

Summarizing, we have

$$\alpha \text{ is constructible} \Leftrightarrow \alpha \in \mathscr{C}^{\surd} \Leftrightarrow G(Q(\alpha)/Q) \text{ is constructible.}$$

If we are given a geometric problem, we try first to transform it into the problem of constructing the roots of some polynomial $f(x) \in Q[x]$. Next we try to determine the group of $f(x)$. For example, if we wish to construct a cube of volume 2, we must construct a side of length $\sqrt[3]{2}$. Therefore, we must construct the root $\alpha = \sqrt[3]{2}$ of the polynomial $f(x) = x^3 - 2$. The group of this is $\mathfrak{S}_3$, whose composition factors have orders 2 and 3. The group $\mathfrak{S}_3$ is therefore not constructible, so $\sqrt[3]{2}$ is not, either, leaving the duplication of the cube an unsolvable problem.

We saw earlier that a polynomial has either none or all its roots expressible in radicals. We now have the corresponding result for constructible roots: Either none or all the roots of $f(x) \in Q[x]$ are constructible, because the constructibility of a root depends only on the group of the equation.

Construction of a Regular Polygon of n Sides.   Since the nth roots of unity are $\varepsilon = \cos(2\pi k/n) + i\sin(2\pi k/n)$, they form the vertices of a regular polygon inscribed in the unit circle. Asked to inscribe a regular polygon in a given circle, we consider it as the unit circle and try to construct the nth roots of unity. The Galois group G_ε of the nth roots is of order $\varphi(n)$, where $\varphi(n)$ is the Euler φ-function, because $\deg \Phi_n(x) = \varphi(n)$, and it is always Abelian. The orders of its composition factors are the primes p which divide $\varphi(n)$. Since G_ε is constructible if and only if these primes are all equal to 2, we must have the following theorem:

THEOREM.   A regular polygon of n sides is constructible $\Leftrightarrow \varphi(n) = 2^k$ for some integer k.

4.6 RULER-AND-COMPASS CONSTRUCTIONS

It can be shown very easily that $\varphi(n) = 2^k$ if and only if $n = 2^r \cdot p_1 \cdot p_2 \cdot \ldots \cdot p_3$, where $p_1, \ldots, p_3$ are s different primes, each of the form $(2^{2m} + 1)$. In particular, we have

$$\varphi(n) = n \prod_{p \mid n} \left(1 - \frac{1}{p}\right).$$

THEOREM. If p is a prime, then the regular polygon of p sides is constructible $\Leftrightarrow p$ is of the form $p = 2^{2^m} + 1$.

Proof. If p is a prime, then the number of integers less than p and prime to p is

$$\varphi(p) = p - 1.$$

Therefore the regular p-gon is constructible if and only if

$$p - 1 = 2^r, \quad \text{for some integer } r.$$

Now suppose we have r factors, say $r = 2^k(2s + 1)$. If we write $t = 2^{2^k}$, then

$$p = 2^r + 1 = 2^{2^k(2s+1)} + 1 = (2^{2^k})^{2s+1} + 1 = t^{2s+1} + 1$$

$$= (t + 1)(t^{2s} - t^{2s-1} + t^{2s-2} - \cdots + 1),$$

so unless $s \neq 0$, p is not prime. If p then is to be a prime, s must thus be 0. ‖

The smallest primes of this form are $p = 2^{2^1} + 1 = 5$, $p = 2^{2^2} + 1 + 17$, and $p = 2^{2^3} + 1 = 257$. We have already constructed a pentagon and the amount of work involved in constructing a 257-gon is well beyond the scope of any book, so we shall show how to construct a regular 17-gon. Most people who have actually carried out this construction agree it was fun to see it work, but doing it once in a lifetime is quite enough.

We must solve the polynomial

$$\Phi_{17}(x) = x^{16} + x^{15} + x^{14} + \cdots + x + 1,$$

whose roots are $\varepsilon, \varepsilon^2, \ldots, \varepsilon^{16}$, where $\varepsilon^{17} = 1$ and ε is a primitive seventeenth root.

The group G_ε of this polynomial is isomorphic to $\mathbf{Z}_{17}^\times$, which is cyclic with 16 elements. Every $\tau \in G_\varepsilon$ is determined by a relation of the form $\tau(\varepsilon) = \varepsilon^k$ for some k, $1 \le k \le 16$. As we saw earlier,

$$\tau(\varepsilon^i) = (\tau\varepsilon)^i = (\varepsilon^k)^i = \varepsilon^{ki},$$

$$\tau^r(\varepsilon) = \tau(\tau \cdots \tau(\varepsilon)) = \varepsilon^{kr}.$$

Since G_ε is cyclic there is some element σ in G_ε such that

$$G_\varepsilon = \{(1), \sigma, \sigma^2, \ldots, \sigma^{15}\}$$

with $\sigma^{16} = (1)$, $\sigma^r \ne (1)$ for $1 \le r < 16$, where (1) is the identity automorphism: $(1)(\varepsilon) = \varepsilon$. We must therefore find an integer k such that

$$\sigma(\varepsilon) = \varepsilon^k,$$

$$\sigma^r(\varepsilon) \ne (1)(\varepsilon), \qquad \text{that is, } \sigma^r(\varepsilon) = \varepsilon^{k^r} \ne \varepsilon, \text{ for } 1 \le r < 16,$$

$$\sigma^{16}(\varepsilon) = (1)(\varepsilon), \qquad \text{that is, } \sigma^{16}(\varepsilon) = \varepsilon^{k^{16}} = \varepsilon.$$

We know that $\varepsilon^{17} = 1$, so k must satisfy the conditions

$$k^r \not\equiv 1 \ (\text{mod } 17), \qquad \text{for } 1 \le r < 16,$$

$$k^{16} \equiv 1 \ (\text{mod } 17).$$

This means that k must be a primitive sixteenth root of 1 (mod 17). By trial and error we find that $k = 3$ satisfies these conditions, but there are other possible choices, for example, $k = $ _____. If we take $k = 3$, we get the following table for σ:

ε^i	ε	ε^2	ε^3													
$\sigma(\varepsilon^i)$	ε^3	ε^6	ε^9													

4.6 RULER-AND-COMPASS CONSTRUCTIONS

The automorphism σ therefore causes the following permutation of the roots:

$$\varepsilon \to \varepsilon^3 \to \varepsilon^9 \to \varepsilon^{10} \to \underline{\qquad} \to \underline{\qquad} \to \underline{\qquad} \to \underline{\qquad} \to \underline{\qquad}$$

$$\to \underline{\qquad} \to \underline{\qquad} \to \underline{\qquad} \to \underline{\qquad} \to \underline{\qquad} \to \underline{\qquad} \to \underline{\qquad} \to \underline{\qquad}.$$

This is represented by the permutation

$$\theta = (1\ 3\ 9\ 10\ \underline{\hspace{6cm}}\}$$

of the exponents of ε, and $G_\varepsilon = \{\theta, \theta^2, \ldots, \theta^{16} = (1)\}$ as a permutation group. A composition series for G_ε is

$$G_\varepsilon = G_0 \underset{2}{\rhd} G_1 \underset{2}{\rhd} G_2 \underset{2}{\rhd} G_3 \underset{2}{\rhd} G_4 = (1),$$

where

$$G_0 = \{(1), \sigma, \sigma^2, \ldots, \sigma^{15}\},$$

$$G_1 = \{(1), \sigma^2, \underline{\hspace{5cm}}\},$$

$$G_2 = \{(1), \sigma^4, \underline{\hspace{2.5cm}}\},$$

$$G_3 = \{(1), \underline{\hspace{1.5cm}}\},$$

$$G_4 = (1).$$

The quotient groups are

	Cosets of G_{i+1} in G_i
G_0/G_1	$G_1 = \{(1), \sigma^2, \underline{\hspace{4cm}}\}$, $\sigma G_1 = \{\underline{\hspace{4cm}}\}$
G_1/G_2	$G_2 = \{\underline{\hspace{3cm}}\}, \sigma^2 G_2 = \{\underline{\hspace{2cm}}\}$
G_2/G_3	
G_3/G_4	

We are now ready to start the actual solution of $\Phi_{17}(x) = 0$ and shall describe the construction of a regular 17-gon simultaneously. It can be carried out on a sheet of paper 11 by 16 inches using a scale of 1 unit = 4 inches.

We must construct functions of the roots of $\Phi_{17}(x)$ which are invariant under G_0, G_1, G_2, G_3, respectively. In this particular case we know a great deal about the roots of $\Phi_{17}(x)$: for example, if $\alpha_1 = \varepsilon$ and $\alpha_{16} = \varepsilon^{16}$, that $\alpha_1 \cdot \alpha_{16} = 1$, and the solution will be much easier than it would be for an arbitrary cyclic equation of degree 16.

Step 1. If we now let

$$f_{1,1} = \varepsilon + \sigma^2(\varepsilon) + \sigma^4(\varepsilon) + \cdots + \sigma^{14}(\varepsilon)$$

$$= \underline{\hspace{10cm}},$$

then $f_{1,1}$ is invariant under G_1, as is its conjugate under G_0,

$$f_{1,2} = \sigma(f_{1,1}) = \sigma(\varepsilon + \sigma^2(\varepsilon) + \sigma^4(\varepsilon) + \cdots + \sigma^{14}(\varepsilon))$$

$$= \sigma(\varepsilon) + \sigma^3(\varepsilon) + \cdots + \sigma^{15}(\varepsilon)$$

$$= \underline{\hspace{10cm}}.$$

Moreover, $(f_{1,1} + f_{1,2})$ and $(f_{1,1} \cdot f_{1,2})$ are both invariant under G_0, so they must be elements of Q. They are not difficult to evaluate using the coefficients of $\Phi_{17}(x)$, for

$$f_{1,1} + f_{1,2} = \underline{\hspace{8cm}}$$

$$= \underline{\hspace{4cm}},$$

$$f_{1,1} \cdot f_{1,2} = (\underline{\hspace{6cm}})$$

$$\cdot (\underline{\hspace{6cm}})$$

$$= \underline{\hspace{8cm}}$$

$$= \underline{\hspace{4cm}}.$$

4.6 RULER-AND-COMPASS CONSTRUCTIONS

Therefore, $f_{1,1}$ and $f_{1,2}$ are the roots of the equation

$$y^2 - (\underline{\hspace{2cm}})y + \underline{\hspace{2cm}} = 0,$$

and their values are $\underline{\hspace{3cm}}$ and $\underline{\hspace{3cm}}$. The decision as to which is $f_{1,1}$ and which $f_{1,2}$ depends on the numbering of the roots. We may suppose that

$$f_{1,1} = \tfrac{1}{2}(-1 + \sqrt{17}) \qquad \text{and} \qquad f_{1,2} = \underline{\hspace{2cm}}.$$

We can now construct segments of lengths $\tfrac{1}{2}(-1 \pm \sqrt{17})$ (remembering that $17 = 4^2 + 1^2$ makes the construction simple).

Step 2. Let

$$f_{2,1} = \varepsilon + \sigma^4(\varepsilon) + \sigma^8(\varepsilon) + \sigma^{12}(\varepsilon) = \underline{\hspace{4cm}},$$

whose conjugate under G_1 is

$$f_{2,2} = \sigma^2(f_{2,1}) = \sigma^2(\varepsilon) + \underline{\hspace{3cm}}$$

$$= \underline{\hspace{3cm}}.$$

These are both invariant under G_2, whereas $(f_{2,1} + f_{2,2})$ and $f_{2,1} \cdot f_{2,2}$ are invariant under G_1. We have

$$f_{2,1} + f_{2,2} = \underline{\hspace{6cm}}$$

$$= \underline{\hspace{3cm}}$$

from step 1, and

$$f_{2,1} \cdot f_{2,2} = (\underline{\hspace{3cm}}) \cdot (\underline{\hspace{3cm}})$$

$$= \underline{\hspace{6cm}}$$

$$= \underline{\hspace{1.5cm}}.$$

So $f_{2,1}$ and $f_{2,2}$ are the roots of the equation

$$z^2 - (\underline{\hspace{4cm}})z + (\underline{\hspace{2cm}}) = 0,$$

and their values are _____ and _____

_____, say

$$f_{2,1} = \frac{-1 + \sqrt{17} + \sqrt{34 - 2\sqrt{17}}}{4} \qquad \text{and}$$

$$f_{2,2} = \underline{\hspace{5cm}}.$$

Similarly if we let

$$f'_{2,1} = \varepsilon^3 + \sigma^4(\varepsilon^3) + \sigma^8(\varepsilon^3) + \sigma^{12}(\varepsilon^3) = \underline{\hspace{5cm}},$$

$$f'_{2,2} = \sigma^2(f'_{2,1}) = \underline{\hspace{4cm}}$$

$$= \underline{\hspace{4cm}},$$

then

$$f'_{2,1} + f'_{2,2} = \varepsilon^3 + \underline{\hspace{8cm}}$$

$$= \underline{\hspace{4cm}},$$

$$f'_{2,1} \cdot f'_{2,2} = \underline{\hspace{4cm}} = \underline{\hspace{4cm}}.$$

So $f'_{2,1}$ and $f'_{2,2}$ are roots of the equation

$$z^2 - (\underline{\hspace{4cm}})z + (\underline{\hspace{2cm}}) = 0,$$

and their values are

$$f'_{2,1} = \underline{\hspace{5cm}} \qquad \text{and}$$

$$f'_{2,2} = \underline{\hspace{5cm}}.$$

4.6   RULER-AND-COMPASS CONSTRUCTIONS

Having already constructed the lengths $\frac{1}{2}(-1 \pm \sqrt{17})$, we can now surely construct segments of length $f_{2,1}$, $f'_{2,1}$, $f_{2,2}$, and $f'_{2,2}$.

Step 3.   Let

$$f_{3,1} = \varepsilon + \sigma^8(\varepsilon) = \underline{\hspace{4cm}},$$

whose conjugate under G_2 is

$$f_{3,2} = \sigma^4(\varepsilon) + \sigma^{12}(\varepsilon) = \underline{\hspace{4cm}}.$$

We have

$$f_{3,1} + f_{3,2} = \underline{\hspace{4cm}} = \underline{\hspace{4cm}},$$

$$f_{3,1} \cdot f_{3,2} = \underline{\hspace{4cm}} = \frac{-1 - \sqrt{17} - \sqrt{34 + 2\sqrt{17}}}{4},$$

so they are the roots of

$$w^2 - (\underline{\hspace{4cm}})w + \underline{\hspace{4cm}} = 0.$$

These roots are

$$f_{3,1} = \underline{\hspace{6cm}},$$

$$f_{3,2} = \underline{\hspace{6cm}}.$$

Since at most square roots are involved, we can construct $f_{3,1}$ and $f_{3,2}$ using, of course, $f_{2,1}$ and $f'_{2,1}$ from above.

Step 4.   Let

$$f_{4,1} = \varepsilon,$$

$$f_{4,2} = \sigma^8(\varepsilon) = \underline{\hspace{3cm}}.$$

We have

$$f_{4,1} + f_{4,2} = \underline{\hspace{4cm}} = \underline{\hspace{4cm}},$$

$$f_{4,1} \cdot f_{4,2} = (\underline{\hspace{2cm}}) \cdot (\underline{\hspace{2cm}}) = \underline{\hspace{3cm}}.$$

So they are the roots of

$$v^2 - (\underline{\hspace{4cm}})v + \underline{\hspace{4cm}} = 0.$$

These roots are

$$f_{4,1} = \underline{\hspace{8cm}} = \varepsilon,$$

$$f_{4,2} = \underline{\hspace{7cm}} = \sigma^8(\varepsilon).$$

Each of these is a primitive seventeenth root. It is a complex number, $\varepsilon = a + ib$, but to construct it it is only necessary to construct a and erect a perpendicular at $(a, 0)$; see Figure 4.4. The intersection of this line with the unit circle will then be ε. Having found ε, complete the 17-gon.

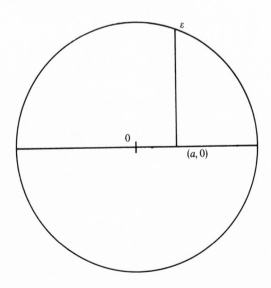

FIGURE 4.4

4.7   LAGRANGE'S THEOREM

The method of solving $\Phi_p(x) = 0$ which is used here dates back to Gauss, although he did not have the group-theoretic motivation. It was a piece of work of which he was particularly proud. The construction of the 17-gon is described by the solution in a completely straightforward manner, but there is a neater, trickier way of doing it [H. W. Richmond, To Construct a Regular Polygon of Seventeen Sides, *Math. Ann.* **67** (1909), p. 459, or H. S. M. Coxeter, *Introduction to Geometry*, Wiley, New York, 1961, p. 27].

EXERCISES.   (1) Use Gauss's method to evaluate the fifth and seventh roots of unity.

(2) *The trisection of angles.* Some angles can be trisected by ruler and compass, for instance an angle of _____°, but not all. Moreover, an angle θ can be constructed if and only if a line of length $\cos \theta$ can also. If we now let $\theta = 30°$, then $\cos \theta =$ _____ and $\theta/3 =$ _____°. From the identity

$$\cos \theta = 4 \cos^3 \frac{\theta}{3} - 3 \cos \frac{\theta}{3}$$

we see that $x = \cos$ _____° must be a root of the equation

$$\underline{\hspace{5cm}}.$$

This equation is irreducible over Q, because _____ _____, so its solution involves cube roots and x is not constructible and neither is an angle of _____°.

4.7   LAGRANGE'S THEOREM

To carry out the solution of a solvable equation we must know the group as a permutation group on the roots, so we need a constructive method for determining these permutations. We should not have to solve the equation to get the group, for that would be circular.

Here and in the following sections we shall show one way in which this can be done. We shall show that given $f(x) \in F[x]$ with roots $\alpha_1, \ldots, \alpha_n$, where char $F = 0$ and $F[x]$ has a factorization algorithm, we can carry out the following steps without first solving $f(x)$:

(1) We can find integers $m_1, \ldots, m_n$ such that if

$$\gamma = m_1\alpha_1 + \cdots + m_n\alpha_n,$$

then (a) $E = F(\gamma)$, and (b) the $n!$ values

$$\sigma_i\gamma = m_1\gamma_{\sigma_i(1)} + \cdots + m_n\gamma_{\sigma_i(n)}$$

where $\sigma_i \in \mathfrak{S}_n$, are all different.

(2) For each σ_i we can construct a different polynomial $p_i(x)$ such that $\sigma_i\gamma = p_i(\gamma)$.

(3) We can find an irreducible polynomial $r(x) \in F[x]$ whose roots are γ and all its conjugates—those $\sigma_i\gamma$ for which σ_i is in the group of $f(x)$.

(4) Lastly, we can then test each $\sigma_i\gamma$ whether it is or is not a root of $r(x)$. Since

$$\sigma_i \text{ is in the group of } f(x) \Leftrightarrow \sigma_i\gamma \text{ is a conjugate of } \gamma$$

$$\Leftrightarrow r(\sigma_i\gamma) = 0$$

$$\Leftrightarrow r(p_i(\gamma)) = 0,$$

this will determine the group of $f(x)$.

For this plan we shall need a theorem due to Lagrange. Although we shall only need special cases of it and could avoid proving the general theorem, but it seems well worth the trouble to derive such a nice result, as it takes very few extra lines. The statement here is a slight variant of the theorem as it was first proved by Lagrange in 1770 [*Nouv. Mem. Acad. Berlin*, (1770); see also Di, p. 173]. Its basic purpose is to show the effect on a given polynomial $p(x_1, \ldots, x_n) \in F[x_1, \ldots, x_n]$ when various

4.7 LAGRANGE'S THEOREM

permutations of the roots $\alpha_1, \ldots, \alpha_n$ of a given equation $f(x) = 0$, with $f(x) \in F[x]$, are put in for $x_1, \ldots, x_n$. Here the symbols $\alpha_1, \ldots, \alpha_n$ may be thought of as names of the roots, denoting specific real numbers, but we do not assume that we know their values. The equation $f(x) = 0$ may well be unsolvable in radicals. This theorem, together with results of Section 4.8, will show that we can determine the effect of various permutations of the roots without first solving the equation. This, in turn, will allow us to find the group of the equation without having solved it—and knowing the group will tell us a method of solution, if there is one.

THEOREM. Let $\mathfrak{S}_n = \{\tau_1, \tau_2, \ldots, \tau_{n!}\}$ be the symmetric group on n elements, and $\alpha_1, \ldots, \alpha_n$ the roots of an irreducible polynomial $f(x) \in F[x]$. Also let $p(x_1, \ldots, x_n)$, $q(x_1, \ldots, x_n)$ be given polynomials in the n variables $x_1, \ldots, x_n$ and with coefficients also in F, and let

$$\tau_i p = p(\alpha_{\tau_i(1)}, \ldots, \alpha_{\tau_i(n)}),$$

$$\tau_i q = q(\alpha_{\tau_i(1)}, \ldots, \alpha_{\tau_i(n)}).$$

Suppose $\tau_i q \neq \tau_j q$ whenever $i \neq j$. Then we can exhibit two polynomials $h(v)$, $k(v) \in F[v]$ such that for every i ($1 \leq i \leq n!$) we have

$$\tau_i p = \frac{h(\tau_i q)}{k(\tau_i q)}.$$

Before proving this, we give an example that will also illustrate the method of proof.

EXAMPLE. Let $n = 2$, so $\mathfrak{S}_2 = \{(1), (12)\}$, and let α_1, α_2 be the roots of $f(x) = x^2 - 2 \in Q[x]$, so that $\alpha_1 + \alpha_2 = 0$ and $\alpha_1 \alpha_2 = -2$. Let

$$p(x_1, x_2) = 2x_1 + x_2,$$

$$q(x_1, x_2) = x_1 - x_2.$$

Then let

$$p_1 = (1)p = p((1)\alpha_1, (1)\alpha_2) = 2\alpha_1 + \alpha_2,$$

$$p_2 = (12)p = p((12)\alpha_1, (12)\alpha_2) = 2\alpha_2 + \alpha_1,$$

$$q_1 = (1)q = \underline{\hspace{2cm}} = \underline{\hspace{2cm}},$$

$$q_2 = (12)q = \underline{\hspace{2cm}} = \underline{\hspace{2cm}}.$$

We now let v be a new variable and set

$$\varphi(v) = (v - q_1)(v - q_2)$$

$$= (v - \underline{\hspace{2cm}})(v - \underline{\hspace{2cm}})$$

$$= v^2 + (\underline{\hspace{2cm}})v + (\underline{\hspace{3cm}})$$

$$= v^2 - ((\alpha_1 + \alpha_2)^2 - 4\alpha_1\alpha_2)$$

$$= v^2 - (-4)(-2)$$

$$= v^2 - 8,$$

and also set

$$h(v) = p_1\left(\frac{\varphi(v)}{(v - q_1)}\right) + p_2\left(\frac{\varphi(v)}{(v - q_2)}\right)$$

$$= p_1(v - q_2) + p_2(v - q_1)$$

$$= (2\alpha_1 + \alpha_2)(v - \alpha_2 + \alpha_1) + (2\alpha_2 + \alpha_1)(v - \alpha_1 + \alpha_2)$$

$$= v(\underline{\hspace{2cm}}) + (\underline{\hspace{3cm}})$$

$$= (-4)(-2) = 8,$$

4.7   LAGRANGE'S THEOREM

so $h(v)$ turns out to be a constant in this particular case. (This is not true in general.) Also let

$$k(v) = \frac{d}{dv}\varphi(v) = 2v.$$

Then Lagrange's theorem says that

$$p_1 = 2\alpha_1 + \alpha_2 = \frac{h(q_1)}{k(q_1)} = \frac{8}{2(\alpha_1 - \alpha_2)},$$

$$p_2 = \underline{\hspace{2cm}} = \frac{h(q_2)}{k(q_2)} = \frac{8}{2(\underline{\hspace{1.5cm}})}.$$

In this particular case we do know the numerical values of α_1 and α_2 and can easily check the result, but it is important to realize that only the symmetric functions $\sigma_1 = \alpha_1 + \alpha_2 = 0$ and $\sigma_2 = \alpha_1\alpha_2 = -2$ were actually used in the calculation.

Proof of the Theorem.   To simplify the notation we write $\tau_i(q) = q_i$, $\tau_i(p) = p_i$ so the conclusion will read $p_i = h(q_i)/k(q_i)$.

(1) Let

$$\varphi(v) = (v - q_1)(v - q_2)(\cdots)(v - q_{n!}) = \prod_{i=1}^{n!} (v - q_i),$$

and let

$$h(v) = p_1\left(\frac{\varphi(v)}{v - q_1}\right) + p_2\left(\frac{\varphi(v)}{v - q_2}\right) + \cdots + p_{n!}\left(\frac{\varphi(v)}{v - q_{n!}}\right)$$

$$= p_1(v - q_2)(\cdots)(v - q_{n!}) + (v - q_1)p_2(v - q_3)(\cdots)(v - q_{n!})$$

$$+ (v - q_1)(v - q_2)p_3(v - q_4)(\cdots)(v - q_{n!}) + \cdots$$

$$+ (v - q_1)(\cdots)(v - q_{n!-1})p_{n!}.$$

(This is reminiscent of the Lagrange interpolation formula.)

(2) Any permutation τ_i will at most permute the order of the factors of $\varphi(v)$ or the summands of $h(v)$ and leave their values unchanged. Both $h(v)$ and $\varphi(v)$ are therefore symmetric in $\alpha_1, \ldots, \alpha_n$.

(3) As a result the coefficients of $\varphi(v)$ and $h(v)$ are symmetric functions of $\alpha_1, \ldots, \alpha_n$ and so are elements of F which can be explicitly calculated from the given coefficients of $f(x)$.

(4) Moreover,

$$h(q_i) = (q_i - q_i)(\cdots)(q_i - q_{i-1})p_i(q_i - q_{i+1})(\cdots)(q_i - q_{n!})$$

and

$$\frac{d}{dv}\varphi(v) = (v - q_2)(\cdots)(v - q_{n!}) + (v - q_1)(v - q_3)(\cdots)(v - q_{n!})$$

$$+(v - q_1)(v - q_2)(v - q_4)(\cdots)(v - q_{n!}) + \cdots$$

$$+(v - q_1)(\cdots)(v - q_{n!-1}).$$

(5) If we now let

$$k(v) = \frac{d}{dv}\varphi(v),$$

then

$$k(q_i) = \left[\frac{d}{dv}\varphi(v)\right]_{v=q_i} = (q_i - q_1)(\cdots)(q_i - q_{i-1})(q_i - q_{i+1})(\cdots)(q_i - q_{n!}).$$

(6) Therefore,

$$h(q_i) = p_i k(q_i).$$

(7) Since $\varphi(v)$ has coefficients in F which we can calculate using symmetric-function theory, we can find $k(v)$ explicitly and $k(v) \in F[v]$.

(8) By hypothesis, $q_i \neq q_j$ for $i \neq j$, so $k(q_i) \neq 0$, and we get

$$p_i = \frac{h(q_i)}{k(q_i)},$$

which proves Lagrange's theorem.   ∥

The polynomial $\varphi(v)$ which is used here has one very important property: If γ is any root of $\varphi(v)$, and q_1 is chosen properly, then $F(\gamma)$ is the splitting field of the given polynomial $f(x)$. This fact will be proved and used in the next sections.

4.8   RESOLVENT OF A POLYNOMIAL

Let E be the splitting field of $f(x) \in F[x]$ and suppose that char $F = 0$. Here, as in almost all the results of this chapter, it would be enough to suppose that E is separable and F is "large enough" so that at some points in the proof when we must choose an element to satisfy a certain condition we can be sure that there are enough elements available to choose from. We shall again need to assume that there is a finite algorithm for factoring polynomials into irreducible factors over F. The field E will be of some finite degree over F and so $E = F(\gamma)$, for some element γ algebraic over F [A, p. 64].

DEFINITION.   If $F(\gamma)$ is the splitting field of the irreducible polynomial $f(x) \in F[x]$, then an irreducible equation $r(x) = 0$ satisfied by γ is called a *resolvent* of $f(x)$ over F.

If the numerical values of the roots $\alpha_1, \ldots, \alpha_n$ of $f(x)$ are at hand, then there is a well-known construction leading to an element γ generating the splitting field and to the resolvent $r(x) = 0$ [A, p. 65]. Our problem, however, is that of constructing $r(x)$ without first solving the equation $f(x) = 0$, for after all we wish to use the resolvent in order to be able to tell whether $f(x) = 0$ is solvable at all.

We shall proceed as follows. Since E is assumed separable and $f(x)$ irreducible, $f(x)$ will have n distinct roots $\alpha, \ldots, \alpha_n$. Note that simply giving names to these roots does not mean that we know their value, or even that they can be expressed in terms of radicals. The calculations will only involve the values of the symmetric functions of the roots, and these we do know, because they are the coefficients of $f(x)$.

If $v(x_1, \ldots, x_n) = m_1 x_1 + \cdots + m_n x_n$, where $m_1, \ldots, m_n \in F$ (and can even be assumed to integers with $m_1 = 0$, $m_2 = 1$ if char $F = 0$ [D, p. 160]) and $\gamma = m_1 \alpha_1 + \cdots + m_n \alpha_n$, then $\gamma \in F(\alpha_1, \ldots, \alpha_n)$; that is, γ is in the splitting field. Therefore,

$$F(\gamma) \leq F(\alpha_1, \ldots, \alpha_n).$$

We would like to have equality in the above relation, but this clearly calls for some restriction on the choice of $m_1, \ldots, m_n$. For example, $m_1 = \cdots = m_n = 0$ evidently implies $\gamma = 0$ and $F(\gamma) = F$, while in general of course $F \lneq F(\alpha_1, \ldots, \alpha_n)$. Now if only —without solving $f(x) = 0$—we could find $m_1, \ldots, m_n$ such that $v(x_1, \ldots, x_n)$ takes on $n!$ different values as the $n!$ different permutations of $\alpha_1, \ldots, \alpha_n$ are put in for $x_1, \ldots, x_n$, then $F(\gamma)$ would actually be the splitting field of $f(x)$ over F. The next result shows how this can be accomplished.

THEOREM. If $\alpha_1, \ldots, \alpha_n$ are the roots of the irreducible polynomial $f(x) \in F$, and

$$v(x_1, \ldots, x_n) = m_1 x_1 + \cdots + m_n x_n$$

takes on $n!$ different values as the $n!$ different possible permutations of $\alpha_1, \ldots, \alpha_n$ are substituted for $x_1, \ldots, x_n$, and we let

$$\gamma = m_1 \alpha_1 + \cdots + m_n \alpha_n,$$

then $F(\gamma) = F(\alpha_1, \ldots, \alpha_n)$.

Proof. From the definition of γ it is clear that $F(\gamma) \leq F(\alpha_1, \ldots, \alpha_n)$. We must prove the opposite inclusion. Let j be any integer, $1 \leq j \leq n$. In Lagrange's theorem we set

$$q(x_1, \ldots, x_n) = v(x_1, \ldots, x_n),$$

$$p(x_1, \ldots, x_n) = x_j$$

and let $i = 1$, so $\tau_i = \tau_1 = (1)$. The conclusion of Lagrange's theorem then reads

$$\alpha_j = \frac{h(\gamma)}{k(\gamma)}.$$

Therefore, $\alpha_j \in F(\gamma)$, for every $j = 1, \ldots, n$, and so

$$F(\alpha_1, \ldots, \alpha_n) \leq F(\gamma),$$

which is the result we wanted. ‖

The next problem, therefore, is to pick $m_1, \ldots, m_n \in F$ in such a way that $\tau_i(v_1) \neq \tau_j(v_1)$ whenever $i \neq j$, where we again write v_i for $v(\alpha_{\tau_i(1)}, \ldots, \alpha_{\tau_i(n)})$ and shall prove the following theorem.

THEOREM. It is possible to determine n elements $m_1, \ldots, m_n \in F$ such that $v_i \neq v_j$ whenever $i \neq j$ $(i, j = 1, \ldots, n!)$ without first solving the equation $f(x) = 0$.
Proof. (1) Form $\varphi(v) = \prod_{i=1}^{n!} (v - v_i)$ and let

$$\Delta = \prod_{1 \leq i < j \leq n!} (v_i - v_j)^2$$

be the discriminant of $p(v)$.

(2) $v_i = v_j$ for some $i \neq j$ would mean that the discriminant Δ of $\varphi(v)$ vanishes; that is, $\Delta = 0$.

(3) Δ is however symmetric in $v_1, \ldots, v_{n!}$ and therefore symmetric in $\alpha_1, \ldots, \alpha_n$, and so it can be explicitly calculated from the coefficients of $f(x)$.

(4) The expression obtained in this way will be a polynomial of degree at most $2(n!)$ in $m_1, \ldots, m_n$ with coefficients in F.

(5) We can, therefore, by trial and error, actually find $m_1, \ldots, m_n \in F$ for which $\Delta \neq 0$. Since only a finite number of values must be avoided, we can even find positive integers for which this is true. ‖

The procedure used in the proof of this theorem is quite constructive, although admittedly rather laborious. Note that by construction, $\varphi(v)$ is symmetric in $\alpha_1, \ldots, \alpha_n$, so the coefficients of $\varphi(v)$ can all be calculated from those of $f(x)$, which are known. In practice it is often simplest to guess at $m_1, \ldots, m_n$, evaluate $\varphi(v)$, and check it for double roots by finding the

$$\gcd(\varphi(v), \frac{d}{dv}\varphi(v)).$$

Having now picked $m_1, \ldots, m_n$ so as to make $\Delta \neq 0$, we know that $\gamma = m_1\alpha_1 + \cdots + m_n\alpha_n$ generates the splitting field E over F and that γ is a root of the polynomial $\varphi(v)$ with coefficients in F. If $\varphi(v)$ is irreducible over F, then it satisfies the definition of a resolvent, and we have solved the problem of this section. This is, however, not always the case, and in general we have

$$\varphi(v) = r_1(v) \cdot r_2(v) \cdot \ldots \cdot r_k(v),$$

where $r_1(v), \ldots, r_k(v)$ are irreducible over F, and γ is a root of some one of them.

One of our basic assumptions was the existence of an algorithm that yields a factorization of $\varphi(v)$ in a finite number of steps, so we can find $r_1(v), \ldots, r_k(v)$. But which of these has γ as a root? If it is $r_i(v)$, then r_i satisfies the definition of a resolvent, and we are done, provided we can recognize the correct factor $r_i(v)$. We shall show below that this time at least, the answer to our question could hardly be more convenient: $r_1, \ldots, r_k$ all have the same degree and any one of them is a resolvent, so no further calculation is necessary:

THEOREM.   Given the irreducible polynomial $f(x) \in F[x]$ with splitting field E. We can form $\varphi(v) \in F[v]$ as above. Then

(1) All the irreducible factors of $\varphi(v)$ have the same degree.

(2) Whenever δ is the root of some irreducible factor $r(v)$ of $\varphi(v)$, then $F(\delta) = E$.

("Irreducible," of course, means "irreducible over F.")

Proof.   We shall again use Lagrange's theorem:

(1) Let δ be any root of $p(v)$, so δ is a root of some irreducible factor $r(v)$.

4.8 RESOLVENT OF A POLYNOMIAL

(2) By the construction of $p(v)$ we know that there must be some permutation τ_k such that $\delta = v_k$. (For the roots of $p(v)$ are just $v_1, \ldots, v_{n!}$.)

(3) Apply Lagrange's theorem with

$$q(x_1, \ldots, x_n) = m_1 x_1 + \cdots + m_n x_n,$$

$$p(x_1, \ldots, x_n) = m_1 x_{\tau_k(1)} + \cdots + m_n x_{\tau_k(n)}$$

and let the permutation used in the conclusion of Lagrange's theorem be the identity permutation (1).

(4) We get

$$\delta = v_k = \frac{h(v_1)}{k(v_1)}.$$

(5) We already know that $F(\gamma) = E$, where $\gamma = v_1$.

(6) By (4), we see that $\delta \in F(\gamma)$, hence $\delta \in E$ and $F(\delta) \leq E$.

(7) To get the converse inclusion, we again use Lagrange's theorem, now taking

$$q(x_1, \ldots, x_n) = m_1 x_1 + \cdots + m_n x_n,$$

$$p(x_1, \ldots, x_n) = m_1 x_{\tau_j(1)} + \cdots + m_n x_{\tau_j(n)},$$

where j is picked so that $\tau_k \tau_j = (1)$, that is, $\tau_j = \tau_k^{-1}$.

(8) We get

$$\tau_k(p) = \frac{h(\tau_k(q))}{k(\tau_k(q))},$$

with $\tau_k(q) = v_k$ and $\tau_k(p) = \tau_k(\tau_j(v_1)) = \tau_k(\tau_k^{-1}(v_1))$, so that

$$\gamma = v_1 = \tau_k(\tau_k^{-1} v_1) = \tau_k(v_j) = \frac{h(v_k)}{k(v_k)} = \frac{h(\delta)}{k(\delta)}.$$

(9) So $\gamma \in F(\delta)$.

(10) From (6) and (9) we get $F(\delta) = E$ whenever δ is a root of an irreducible factor $r(v)$ of $p(v)$.

(11) Let δ_1 be a root of the factor $r_1(v)$ and δ_2 a root of $r_2(v)$.

(12) Then

$$\deg r_1(v) = [F(\delta_1):F] = [E:F] = [F(\delta_2):F] = \deg r_2(v),$$

which completes the proof of the theorem. $\|$

In order to find a resolvent of $f(x)$ we therefore simply form $\varphi(v)$, carefully choosing $m_1, \ldots, m_n$, so that $\varphi(v)$ has no double roots, and then let $r(v)$ be any irreducible factor of $p(v)$. The polynomial $\varphi(v)$ is often called a *complete Galois resolvent* of $f(x)$ over F. We then have the result that any irreducible factor of a complete Galois resolvent of $f(x)$ over F is a resolvent.

EXAMPLE.   Let

$$f(x) = x^3 - 2,$$

$$F_1 = Q,$$

$$F_2 = Q(\omega),$$

and let us find a resolvent for $f(x)$ over F_1 and over F_2 by this method. In this example it is possible to tell on inspection that $(x^3 - 2)$ itself is a resolvent over $Q(\omega)$. A resolvent over Q is a little less evident. We shall, however, need, and so shall compute, the particular form for the resolvent given by the theorems we just proved with known values of $m_1, \ldots, m_n$, in order to calculate the Galois group of $f(x)$ in Section 4.9. Of course we already know the group of this particular polynomial, but the point of the example is that it is possible to calculate the resolvent and the group using only the coefficients of $f(x)$ by a method that works in general. In the expression for v_1, we can always take $m_1 = 0$, $m_2 = 1$ [Di, p. 161]. We shall try $m_3 = -1$ and then

4.8 RESOLVENT OF A POLYNOMIAL

check $\varphi(v)$ for multiple roots. Therefore, we have $\mathfrak{S}_3 = \{\tau_1, \ldots, \tau_6\}$, where the permutations are numbered

τ_1	τ_2	τ_3	τ_4	τ_5	τ_6
(1)	(12)	(13)	(23)	(123)	(132)

Then

$$v_1 = m_1\alpha_1 + m_2\alpha_2 + m_3\alpha_3$$

$$= \alpha_2 - \alpha_3,$$

$$v_2 = \alpha_1 - \alpha_3 = (12)v_1,$$

$$v_3 = \underline{\hspace{2cm}} = (13)v_1,$$

$$v_4 = \underline{\hspace{2cm}} = (23)v_1,$$

$$v_5 = \underline{\hspace{2cm}} = (123)v_1,$$

$$v_6 = \underline{\hspace{2cm}} = (132)v_1,$$

$$\varphi(v) = (v - v_1)(v - v_2)(v - v_3)(v - v_4)(v - v_5)(v - v_6)$$

$$= (v - (\alpha_2 - \alpha_3)) \cdot (v - (\alpha_1 - \alpha_3)) \cdot (v - (\alpha_2 - \alpha_1)) \cdot (v - (\alpha_3 - \alpha_2))$$

$$\cdot (v - (\alpha_1 - \alpha_2)) \cdot (v - (\alpha_3 - \alpha_1))$$

$$= (v - (\alpha_2 - \alpha_3)) \cdot (v + (\alpha_2 - \alpha_3)) \cdot (v - (\alpha_1 - \alpha_3)) \cdot (v + \underline{\hspace{2cm}})$$

$$\cdot (v - \underline{\hspace{2cm}}) \cdot (v + \underline{\hspace{2cm}})$$

$$= (v^2 - (\alpha_2 - \alpha_3)^2) \cdot (\underline{\hspace{3cm}})$$

$$\cdot (\underline{\hspace{3cm}})$$

$$= v^6 - v^4((\alpha_2 - \alpha_3)^2 + \underline{\hspace{2cm}} + \underline{\hspace{2cm}})$$

$$+ v^2((\alpha_2 - \alpha_3)^2(\alpha_1 - \alpha_3)^2 + \underline{\hspace{3cm}}$$

$$+ \underline{\hspace{3cm}}) - ((\alpha_2 - \alpha_3)^2(\alpha_1 - \alpha_3)^2(\alpha_1 - \alpha_2)^2).$$

The coefficients are all symmetric in α_1, α_2, α_3 and must be evaluated using only the elementary symmetric functions. Since $f(x) = x^3 - 2$, we have

$$\sigma_1 = \alpha_1 + \alpha_2 + \alpha_3 = \underline{\hspace{1cm}},$$

$$\sigma_2 = \alpha_1\alpha_2 + \alpha_2\alpha_3 + \alpha_3\alpha_1 = \underline{\hspace{1cm}},$$

$$\sigma_3 = \alpha_1\alpha_2\alpha_3 = 2.$$

Using Newton's identities (Section 1.5) we get

$$s_2 = \sum \alpha_1^2 = \underline{\hspace{3cm}}$$

and calculate the symmetric functions

$$\sum (\alpha_i - \alpha_j)^2 = 2s_2 - 2\sigma_2 = \underline{\hspace{3cm}},$$

$$\sum (\alpha_i - \alpha_j)^2(\alpha_j - \alpha_k)^2 = 2\sigma_2^2 + \sigma_2^2 - 2s_2\sigma_2 = \underline{\hspace{3cm}}.$$

The constant term of $\varphi(v)$ is the discriminant Δ of $f(x)$,

$$\Delta = (\alpha_2 - \alpha_1)^2(\alpha_1 - \alpha_3)^2(\alpha_3 - \alpha_2)^2.$$

The value of Δ can also be found using symmetric functions and we get

$$\Delta = (\alpha - \alpha_2)^2(\alpha_2 - \alpha_3)^2(\alpha_3 - \alpha_1)^2$$

$$= s_2 \cdot \sigma_2^2 - 27\sigma_3^2 + 14\sigma_1\sigma_2\sigma_3 - 2\sigma_1s_2\sigma_3$$

$$- 2\sigma_1^3\sigma_3 - 2\sigma_2^3$$

$$= \underline{\hspace{3cm}} = -108.$$

4.8 RESOLVENT OF A POLYNOMIAL

Therefore,

$$\varphi(v) = x^6 + 108.$$

This has no repeated roots, because _____

_____, so the choice of $m_3 = -1$ was a good one. Over $F_1 = Q$, this polynomial is irreducible, so it is itself a resolvent; over $F_2 = Q(\omega)$ it factors

$$\varphi(v) = (x^3 - 6\sqrt{-3})(x^3 + \sqrt[6]{-3}),$$

and each of the two factors is a resolvent. (It is easy to check here that this result is correct and we really do have $E = Q(\omega, \sqrt[3]{2}) = Q(\gamma)$, where γ is a root of $(x^6 + 108)$, for $\gamma = \sqrt[6]{-108} = \sqrt{-3\sqrt[3]{2}} = (-1 - 2\omega) \cdot \sqrt[3]{2} \in Q(\omega, \sqrt[3]{2})$, and

$$\omega = -\tfrac{1}{2} - \tfrac{1}{2}\sqrt{-3} = -\tfrac{1}{2} - \tfrac{1}{2}(-\tfrac{1}{6}\gamma^3) = -\tfrac{1}{2} + \tfrac{1}{12}\gamma^3, \qquad \sqrt[3]{2} = \tfrac{1}{18}\gamma^4,$$

so $\omega, \sqrt[3]{2} \in Q(\gamma)$.)

Notice that we verified that $[E\!:\!F] = \deg r(x)$, for over F_1 the polynomial $x^6 + 108$ is irreducible of degree 6, while over F_2 the degree of the resolvent is 3 and indeed $[Q(\omega, \sqrt[3]{2})\!:\!Q(\omega)] = 3$. If we wanted to find a resolvent over the field $Q(\sqrt[3]{2})$ we would have $[Q(\omega, \sqrt[3]{2})\!:\!Q(\sqrt[3]{2})] = $ _____, so $x^6 + 108$ must factor over $Q(\sqrt[3]{2})$ into _____ factors, each of degree _____. We get

$$x^6 + 108 = (\underline{\hspace{3cm}})(x^2 + 3\sqrt[3]{2}x + 3 \cdot 2^{2/3})(\underline{\hspace{3cm}})$$

over $Q(\sqrt[3]{2})$. Each factor is a resolvent of $f(x)$ over $Q(\sqrt[3]{2})$.

Incidentally, it might be worth noting that the construction of the resolvent already shows how to determine the order of the Galois group of an equation: For we can certainly form $\varphi(v)$, and our basic hypothesis is that there is an algorithm for finding the irreducible factor of polynomials over F, so we can factor $\varphi(v)$ and find a resolvent. The fundamental theorem of Galois theory then tells that $|G(E/F)| = [E\!:\!F] = [F(v_1)\!:\!F] = \deg r_1(v)$. If $\deg f(x) = 3$, this is sufficient to determine the Galois group, as the only transitive subgroups of $\mathfrak{S}_3$ are $\mathfrak{A}_3$ and $\mathfrak{S}_3$ itself. If deg $f(x) = 4$, knowing the order of the Galois group is almost enough; for closer examination of $\mathfrak{S}_4$ shows that there are only five transitive subgroups of $\mathfrak{S}_4$ and these are groups of orders 24, 12, 8, and two groups of order 4: Klein's 4-group $\mathfrak{B}$ and the cyclic group $\mathscr{C}_4$. Hence to determine the group of a quartic, it is further only necessary to distinguish between the last two. This can be done as follows: $G(E/F)$ is a subgroup of the alternating group $\mathfrak{A}_n$ if the discriminant of $f(x)$ is a perfect square. (Why?) The discriminant of a quartic can surely be found. The group is then $\mathfrak{B}$ if the resolvent is of degree 4 and the discriminant is a square, and it is $\mathscr{C}_4$ if the resolvent is of degree 4 and the discriminant is not a square. To determine the group of a quartic, it is therefore sufficient to find the degree of the resolvent and the discriminant. Each of these can be calculated. If $f(x)$ is of degree 5 or 7, the possible Galois groups are again uniquely determined by their orders, so it is only necessary to find the degree of the resolvent in these cases.

We can use the construction of the resolvent to show that there is an algorithm (decision method) for determining whether $E_1 = E_2$, where E_1 and E_2 are the splitting fields of $p_1(x)$ and $p_2(x)$. If $p_1(x)$ and $p_2(x)$ are quadratic, the problem is trivial, but in general, of course, the polynomials $p_1(x)$ and $p_2(x)$ need not even be solvable. Nonetheless, we can form the resolvents $r_1(x)$ of $p_1(x)$, $r_2(x)$ of $p_2(x)$, and $r_3(x)$ of the product $p_1(x) \cdot p_2(x)$ and conclude that $E_1 = E_2$ if $\deg r_1(x) = \deg r_2(x) = \deg r_3(x)$.

Exercises. (1) Let α be a root of $p_1(x)$ and β of $p_2(x)$, where $p_1(x)$ and $p_2(x)$ are irreducible over Q. Show that we can find $n \in Q$ such that $Q(\alpha, \beta) = Q(\gamma)$, where $\gamma = \alpha + n\beta$. (*Hint:* See MacD, p. 96. Notice how this proof is related to Lagrange's theorem.)

(2) Given $p(\alpha) = p(\beta) = 0$. Describe an algorithm for determining whether $Q(\alpha) = Q(\beta)$. (*Hint:* Let $\gamma = \alpha + n\beta$ and compare $[Q(\gamma):Q]$ with $[Q(\alpha):Q]$ and $[Q(\beta):Q]$.)

(3) Let α be a root of the irreducible polynomial $p(x)$ with coefficients in Q. Describe an algorithm for deciding whether a polynomial $q(x)$ is reducible or irreducible over the field $Q(\alpha)$.

(4) Use Exercises 2 and 3 to describe an algorithm that will decide whether a given algebraic number β is in the field $Q(\alpha)$.

4.9 CALCULATION OF THE GALOIS GROUP

We are now ready to show how one can find the Galois group of an irreducible polynomial $f(x)$ without first finding its roots. To do this, we choose suitable values for $m_1, m_2, \ldots, m_n$ and then form $\varphi(v)$ as in Section 4.8. We now pick $r(v)$ to be one of the irreducible factors of $\varphi(v)$ over F and we may assume that the roots $\alpha_1, \ldots, \alpha_n$ are labeled in such a way that $\gamma = v_1 = m_1\alpha_1 + \cdots + m_n\alpha_n$ is a root of $r(v) = 0$. If $\deg r(v) = v$, let $v_{j_1}, \ldots, v_{j_v}$ be the v roots of $r(v)$. Since v_1 is among these, we may suppose that it is v_{j_1}, so $j_1 = 1$. Each irreducible factor of $\varphi(v)$ is a resolvent, so every root v_i of $\varphi(v)$ must generate the splitting field E, that is, $F[v_i] = F[\gamma] = E$ for every i, j between 1 and $n!$. In particular, every $v_i \in F[\gamma]$, and therefore each v_i, is a polynomial $l_i(\gamma)$. Can we say any more without solving $f(x)$? The next theorem says that we can; in fact, we can express each v_i explicitly as a polynomial in γ without first solving $f(x) = 0$.

THEOREM. Let the resolvent $r(x)$ of the irreducible polynomial $f(x)$ be formed as in Section 4.8. Then we can find $n!$ polynomials $l_1(v), \ldots, l_{n!}(v)$ such that $v_j = l_j(v_1)$, for every $j = 1, \ldots, n!$.

Proof. Again we use Lagrange's theorem, letting

$$p(x_1, \ldots, x_n) = \tau_j(m_1 x_1 + \cdots + m_n x_n),$$

and $q(x_1, \ldots, x_n) = m_1 x_1 + \cdots + m_n x_n$. We can, therefore, find two polynomials $h_j(v)$ and $k_j(v)$ with coefficients in F such that

$$(1)(\tau_j(m_1\alpha_1 + \cdots + m_n\alpha_n)) = \frac{h_j((1)(m_1\alpha_1 + \cdots + m_n\alpha_n))}{k_j((1)(m_1\alpha_1 + \cdots + m_n\alpha_n))}.$$

By definition of $v_1, \ldots, v_{n!}$ (in Section 4.8) we have, therefore,

$$v_j = \frac{h_j(\gamma)}{k_j(\gamma)}.$$

However, γ is a root of $r(v)$, so we can find a polynomial $k'_j(\gamma)$ with coefficients in F such that

$$k_j(\gamma) \cdot k'_j(\gamma) = 1,$$

$$\frac{1}{k_j(\gamma)} = k'_j(\gamma).$$

Substituting this in and letting $l_j(\gamma) = h_j(\gamma) \cdot k'_j(\gamma)$, we get

$$v_j = \frac{h_j(\gamma)}{k_j(\gamma)} = h_j(\gamma) \cdot k'_j(\gamma) = l_j(\gamma). \qquad \|$$

EXAMPLE. Let $F = Q(\omega), f(x) = x^3 - 2$. From the preceding example we know that $r(v) = v^3 + 6\sqrt{-3}$ is a resolvent and that

$$\gamma = v_1 = \alpha_2 - \alpha_3$$

$$v_2 = \alpha_1 - \alpha_3$$

$$v_3 = \alpha_2 - \alpha_1$$

$$v_4 = \alpha_3 - \alpha_2$$

$$v_5 = \alpha_1 - \alpha_2$$

$$v_6 = \alpha_3 - \alpha_1.$$

We want to find $l_1(\gamma), \ldots, l_6(\gamma)$ such that $v_j = l_j(\gamma)$. Obviously $v_1 = \gamma$ and $v_4 = -\gamma$, so $l_1(\gamma) = \gamma$, $l_2(\gamma) = -\gamma$. Here the original equation and the resolvent are both very simple equations and the polynomials $l_i(\gamma)$ could easily be guessed, but we shall go

4.9 CALCULATION OF THE GALOIS GROUP

through the general procedure to find $l_2(\gamma)$ as an illustration. We have (see p. 200), $\tau_2 = (12)$, $v_2 = \tau_2(v_1) = \tau_2(\alpha_2 - \alpha_3) = \alpha_1 - \alpha_3$, and we let $p_1 = \alpha_1 - \alpha_3$,

$p_2 = \tau_2(p_1) = $ _____ , $p_3 = \tau_3(p_1) = $ _____ ,

$p_4 = \tau_4(p_1) = $ _____ , $p_5 = \tau_5(p_1) = $ _____ ,

$p_6 = \tau_6(p_1) = $ _____ ,   and $q_i = v_i$.

Referring back to the proof of Lagrange's theorem we see that we must take

$$h_2(v) = p_1(v - q_2) \cdots (v - q_6)$$

$$+ (v - q_1)p_2(v - q_3) \cdots (v - q_6)$$

$$+ (v - q_1)(v - q_2)p_3(v - q_4)(v - q_5)(v - q_6)$$

$$+ (v - q_1)(v - q_2)(v - q_3)p_4(v - q_5)(v - q_6)$$

$$+ (v - q_1) \cdots (v - q_4)p_5(v - q_6)$$

$$+ (v - q_1) \cdots (v - q_5)p_6.$$

After considerable calculation involving symmetric functions [Section 1.5 or W, p. 105] we see that this gives

$$h_2(v) = v^5(\underline{\hspace{1.5cm}})$$

$$+ v^4(2s_1 - 2\sigma_2)$$

$$+ v^3(-5s_3 + 3\sigma_1 s_2 - 12\sigma_3)$$

$$+ v^2(4\sigma_1 s_3 - 3s_2^2 - 3s_4)$$

$$+ v(5s_5 - 5\sigma_1 s_4 - 12\sigma_3\sigma_2 + 2s_2 s_3 + 15\sigma_3 s^2)$$

$$+ (-6(s_3^2 - s_6) - 27\sigma_3^2 + \tfrac{3}{2}\sigma_1^2(s_2^2 - s_4) - 6\sigma_3 s_3)$$

$$= -54v^3 - 324,$$

when we use Newton's identities and the values of the symmetric functions of α_1, α_2, α_3 obtained from the coefficients of $f(x) = x^3 - 2$.

Since v_1 is a root of $r(v) = v_3 + 6\sqrt{-3}$, we see that $v_1^3 = -6\sqrt{-3}$ and

$$h_2(v_1) = -54(-6\sqrt{-3}) - 324 = (\underline{\hspace{3cm}})\omega.$$

Also

$$k_2(v) = \frac{d}{dv}\varphi(v) = \frac{d}{dv}(v^6 + 108) = 6v^5,$$

so

$$k_2(v_1) = 6v_1^5 = 6v_1^3 \cdot v_1^2 = 6(\underline{\hspace{3cm}})v_1^2.$$

Therefore,

$$v_2 = \frac{h_2(v_1)}{k_2(v_1)} = \underline{\hspace{3cm}} = \frac{18\omega}{\sqrt{-3v_1^2}} = \frac{18\omega}{\sqrt{-3\gamma^2}}.$$

However, $v_1^3 = \gamma^3 = -6\sqrt{-3}$, so we get

$$\gamma^2 = \frac{-6\sqrt{-3}}{\gamma},$$

$$v_2 = \frac{(18\omega)(\gamma)}{(\sqrt{-3})(-6\sqrt{-3})} = \underline{\hspace{3cm}},$$

$$v_6 = -v_2 = \underline{\hspace{3cm}}.$$

One can also show that $v_3 = -v_5 = -\omega^2\gamma$.

Remember now that $\mathfrak{S}_n = \{\tau_1 = (1), \tau_2, \ldots, \tau_{n!}\}$ and $v_k = \tau_k(v_1) = m_1\alpha_{\tau_{k(1)}}$ $+ \cdots + m_n\alpha_{\tau_{k(n)}}$ and let $G = \{\tau_1, \tau_{j_2}, \ldots, \tau_{j_v}\}$. The next theorem shows that not only is G a group, but it is isomorphic to the Galois group of $f(x)$ over F, so it is precisely the group we are trying to determine:

4.9 CALCULATION OF THE GALOIS GROUP

THEOREM. If $r(v)$ is an irreducible factor of $\varphi(v)$ with roots $v_1 = v_{j_1}, v_{j_2}, \ldots, v_{j_\nu}$ and $G = \{\tau_{j_1}, \ldots, \tau_{j_\nu}\}$, then

$$G \cong G(E/F),$$

where E is the splitting field of the irreducible polynomial $f(x) \in F[x]$.

Proof. We shall prove this in three steps. In step (1) we show that for each automorphism θ in $G(E/F)$ there is a unique permutation $\tau \in G$ such that $\theta(v_1) = \tau(v_1)$; in step (2) we show that if θ and ψ are both in $G(E/F)$ and $\theta(v_1) = \psi(v_1)$, then $\theta = \psi$. Steps (1) and (2) together establish a one-to-one correspondence between G and $G(E/F)$. In step (3) we show that this correspondence preserves the group operation.

(1) Let θ be in $G(E/F)$. Since θ is an automorphism of E which leaves every element of F fixed, and $r(v_1) = 0$ by hypothesis, we know that

$$r(\theta(v_1)) = \theta(r(v_1)) = \theta(0) = 0.$$

Therefore, $\theta(v_1)$ is also a root of $r(v) = 0$. But all these roots are of the form $\tau(v)$ with $\tau \in G$, so $\theta(v_i) = v_\tau$ for some $\tau \in G$.

(2) Suppose θ, $\psi \in G(E/F)$ and $\theta(v_1) = v_\tau = \psi(v_1)$. Then, remembering that $F(v_1) = F(v_\tau) = E$, we know that any $\beta \in E$ is a polynomial in v_1 with coefficients in F, say $\beta = P(v_1)$. Therefore, for any $\beta \in E$,

$$\theta(\beta) = \theta(P(v_1)) = P(\theta(v_1)) = P(v_\tau) = P(\psi(v_1)) = \psi P(v_1)) = \psi(\beta);$$

in other words, $\theta = \psi$.

(3) Now suppose that $\theta_j, \theta_k \in G(E/F)$ and that $\theta_j(v_1) = v_j = \tau_j(v_1), \theta_k(v_1) = v_k = \tau_k(v_1)$ are both roots of $r(v)$. By the definition of G, the permutations τ_j and τ_k are therefore in the set G. We must show that $\tau_j \tau_k$ is also in G and that $(\theta_j \theta_k)v_1 = (\tau_j \tau_k)v_1$.

To see that $\tau_j \tau_k \in G$, we show that $(\tau_j \tau_k)v_1$ is also a root of $r(v)$: Since the coefficients of $r(v)$ are all in F and therefore left fixed by τ_j and τ_k, have $r((\tau_j \tau_k)(v_1)) = (\tau_j \tau_k)(r(v_1)) = (\tau_j \tau_k)(0) = 0$, so $\tau_j \tau_k \in G$.

Next we show that the group operation is preserved:

$$(\theta_j\theta_k)v_1 = \theta_j(\theta_k(v_1))$$

$$= \theta_j(v_k) \qquad \text{by definition of } v_k,$$

$$= \theta_j(l_k(v_1)) \qquad \text{by the preceding theorem,}$$

$$= l_k(\theta_j(v_1)) \qquad \text{since } l_k \text{ has coefficients in } F$$
$$\text{and these are left fixed by } \theta_j,$$

$$= l_k(v_j)$$

$$= l_k(\tau_j(v_1)) \qquad \text{by definition of } v_j,$$

$$= \tau_j(l_k(v_1)) \qquad \text{since } l_k \text{ has coefficients in } F$$
$$\text{and these are not affected by } \tau_j,$$

$$= \tau_j(\tau_k(v_1))$$

$$= (\tau_j\tau_k)(v_1). \qquad \|$$

COROLLARY. $G(E/F)$ can be calculated.

Proof. By checking each permutation $\tau_j \in \mathfrak{S}_n$, we can determine exactly which permutations are elements of $G(E/F)$. To do this, we first express every $v_j = \tau_j(v_1)$ as a polynomial $l_j(\gamma)$ and check whether $r(\gamma) = 0$ implies $r(l_j(\gamma)) = 0$. $\quad \|$

EXAMPLE. We continue to take

$$f(x) = x^3 - 2,$$

$$F = Q(\omega),$$

and as the resolvent

$$r(v) = v^3 + 6\sqrt{-3}.$$

4.9 CALCULATION OF THE GALOIS GROUP

From the preceding examples we have

	1	2	3	4	5	6
j	(1)	(12)	(13)	(23)	(123)	(132)
$l_j(\gamma)$	γ		$-\omega^2\gamma$			
$r(l_j(\gamma))$	$\gamma^3 + 6\sqrt{-3} = 0$		$-\gamma^3 + 6\sqrt{-3} \neq 0$			

Therefore, the roots of $r(v)$ are $v_1,$ _____, and _____, and the Galois group of $(x^3 - 2)$ over $Q(\omega)$ is

$$G(E/F) = \{\text{\underline{\hspace{4cm}}}\}.$$

There are other methods of determining the Galois group of a polynomial due to Zassenhaus [dittoed notes, University of Ohio, 1966], and in many cases the method described in [vdW, p. 189] works very well. [See also Section 4.11.]

Given any polynomial $p(x) \in Q[x]$ we can now determine its group. How about the converse problem: Given a group G, is there a polynomial $p(x)$ whose group is just G? If the ground field F is not specified, the answer is yes: For given G of order n we simply consider a polynomial $p(x)$ of degree n whose group is $\mathfrak{S}_n$. Let E be its splitting field. There is a transitive subgroup $G' < \mathfrak{S}_n$ such that $G' \cong G$. Then $G(E/F_{G'}) = G' \cong G$ and $[E:F_{G'}]$ is finite, so the same polynomial $p(x)$ with coefficients considered as elements of $F_{G'}$ has G' as its group over $F_{G'}$. If, however, the ground field is to be Q, then the problem becomes much harder and a general solution is not yet known. In 1954 the Russian mathematician Šafarevich proved that there is such a polynomial $p(x)$ whenever G is solvable [Šafarevič, I. R., On extensions of fields of algebraic numbers solvable in radicals, *Dokl. Akad. Nauk SSSR* (N.S.), **95**, 225–227 (1954)].

If we treat the $m_1, \ldots, m_n$ used in the construction of γ as unknowns, then each $r(l_j(\gamma))$ becomes a polynomial in $m_1, \ldots, m_n$. Suppose some transitive group G on n letters is now given. For each $\sigma_j \in \mathfrak{S}_n$ we can then construct $r(l_j(\gamma))$ and see that there is a polynomial $p(x) \in Q[x]$ with group G if and only if there are integers $m_1, \ldots, m_n$ for which $r(l_j(\gamma)) = 0 \Leftrightarrow \sigma_j \in G'$. This would transform the problem of the existence of a polynomial with group G over Q into an equivalent number-theoretic problem.

4.10 MATRIX SOLUTIONS OF EQUATIONS

This section contains a brief discussion of matrix solutions of polynomial equations. This is prompted by the following well-known theorems:

THEOREM.   If $p(x) \in F[x]$ and C_p is the companion matrix of $p(x)$, then $p(x)$ is the characteristic polynomial of C_p.

THEOREM (Hamilton–Cayley).   Every matrix A satisfies its characteristic polynomial [H, p. 292].

As reminders, we include the following definitions:

DEFINITION.   If A is a matrix, its *characteristic polynomial* $p_A(x)$ is

$$p_A(x) = (-1)^n \det(A - xI).$$

Note that if A is an $n \times n$ matrix, then $\deg p_A(x) = n$ and that $p_A(x)$ is monic.

DEFINITION.   If $p(x) = a_0 + a_1 x + \cdots + a_{n-1} x^{n-1} + x^n$ is a monic polynomial, then its *companion matrix* C_p is

$$C_p = \begin{pmatrix} 0 & 0 \cdots 0 - a_0 \\ 1 & 0 \cdots 0 - a_1 \\ 0 & 1 \cdots 0 - a_2 \\ \cdots\cdots\cdots\cdots \\ 0 & 0 \cdots 1 - a_{n-1} \end{pmatrix}.$$

4.10 MATRIX SOLUTIONS OF EQUATIONS

DEFINITION.   If A is a matrix with elements in F, and $b_0 + \cdots + b_{k-1}x^{k-1} + x^k$ is a monic polynomial of the lowest possible degree having A as a root, then it is called a *minimal polynomial* of A and denoted by $m_A(x)$.

Given A it is easy to show that $m_A(x)$ is uniquely determined, and if $A \neq 0$, then $\deg m_A(x) \geq 1$.

DEFINITION.   Two matrices A and B, both with elements in a field F, are called *similar* (written A sim B) if there is an invertible matrix C such that

$$A = C^{-1}BC.$$

We shall use F_k^k to denote the set of all $k \times k$ matrices with entries in F, and $\mathbf{0}$ will be the zero matrix of whatever dimensions are suited to the context.

Taking the two theorems together shows that every polynomial $p(x)$ has a matrix solution, its companion matrix C_p, so every equation, even the unsolvable polynomial $x^5 - x - 1 = 0$, and any other polynomial unsolvable in radicals, are solvable on inspection in terms of matrices, and that every polynomial is the characteristic polynomial of at least one matrix, namely of C_p. Given any matrix A, it is natural to ask next whether the characteristic equation of A can have any roots other than A itself. In answer to this, we have the following results, where we let $A, B \in F_n^n$, $A, B \neq 0$ and let $p_A(x)$, $p_B(x)$, $m_A(x)$, and $m_B(x)$ be their characteristic and minimal polynomials.

THEOREM.   If $p_A(x)$ is irreducible and separable, and B is a root of $p_A(x)$ (so that $p_A(B) = 0$), then A sim B and the similarity class of A is precisely the set of roots of $p_A(x)$.

Proof.   First notice that if A sim B, then

$$p_A(B) = a_0I + a_1B + \cdots + B^n$$

$$= a_0I + a_1C^{-1}BC + \underline{\hspace{3cm}}$$

$$= C^{-1}(\underline{\hspace{5cm}})C = \mathbf{0}.$$

Therefore, B is a root of $p_A(x)$.

Next we prove two lemmas before concluding the proof of the theorem.

LEMMA 1. If $p_A(B) = 0$ and $p_A(x)$ is irreducible, then $p_A(x) = p_B(x)$; that is, A and B have the same characteristic polynomial.

Proof of Lemma 1. If $p_A(B) = 0$, then $m_B(x)|p_A(x)$, because the minimal polynomial of B is a divisor of every polynomial having B as a root. By hypothesis, $p_A(x)$ is, however, irreducible, and so since $\deg m_B(x) \geq 1$, and both polynomials are monic, we must have $m_B(x) = p_A(x)$. Therefore, $m_B(x)$ is irreducible and of degree n. But we know that $p_B(x)$ is also monic, also of degree n, and that it has $m_B(x)$ as a divisor. This can happen only if $m_B(x) = p_B(x)$. Therefore, $p_B(x) = p_A(x)$. ‖

LEMMA 2. If $p_A(x) = p_B(x)$ and $p_A(x)$ is separable and irreducible, then $A \sim B$.

Proof of Lemma 2. If $p_A(x)$ is separable and irreducible, then its characteristic roots $\lambda_1, \ldots, \lambda_n$ are all different and the Jordan normal form of A is diagonal:

$$
J_A = \begin{pmatrix} \lambda_1 & 0 \cdots 0 \\ \\ 0 & \lambda_1 \cdots 0 \\ \\ \cdots\cdots\cdots\cdots \\ \\ 0 \cdots\cdots\cdots \lambda_n \end{pmatrix}.
$$

By hypothesis, $p_A(x) = p_B(x)$, so of course $J_A = J_B$. By the Jordan normal-form theorem we therefore have $A \sim B$. ‖

Proof of the Theorem. Combine Lemmas 1 and 2. ‖

If $p_A(x)$ is reducible, the characterization of all $n \times n$ matrix roots is a bit more complicated and involves the direct sum of matrices, one for each irreducible factor and sometimes some 1's off the main diagonal—reminiscent of the Jordan normal-form theorem, if there are repeated roots. One may also prove:

THEOREM. If α is a root of the irreducible, separable polynomial $p(x)$ in some algebraic extension of F, then $F(\alpha) \cong F[C_p]$, where $F[C_p]$ is the set of $n \times n$ matrices that are of the form $a_k(C_p)^k + \cdots + a_0 I$, with $a_i \in F$.

This gives a matrix representation for algebraic extensions of F. Can we ask for a little more, say a matrix representation of the splitting field of $p(x)$? If $p(x)$ is of degree n, this may not always be possible in terms of $n \times n$ matrices, but if the Galois group of $p(x)$ is of order k, then we can always do it in terms of $k \times k$ matrices, simply by considering $F[C_{r(x)}]$, where $r(x)$ is a resolvent of $p(x)$. (Remember that the degree of the resolvent equals the order of the Galois group.) We then get the

THEOREM. If E is a splitting field of the irreducible separable polynomial $p(x)$ over F and $r(x)$ is a resolvent of $p(x)$, then

$$E \cong F[C_{r(x)}].$$

Exercises. (1) A given nth-degree irreducible, separable polynomial $p(x) \in F[x]$ may well have an infinite number of roots in F_n^n—all the matrices that are similar to $C_{p(x)}$. In $F[C_{r(x)}]$, however, the polynomial $p(x)$ will have precisely n roots, no two of which are similar. (Show that if A, $B \in F[C_{r(x)}]$ and $A \text{ sim } B$, then $A = B$.)

(2) Discuss the matrix solutions of $p(x)$ in F_k^k, where k is not necessarily equal to the degree of $p(x)$.

4.11 FINITE FIELDS

This section starts with a collection of standard theorems on finite fields. Some of these are very simple and were used earlier, but they are repeated here for completeness. Later in this section there are some applications, as, for instance, the relation between the Galois group of a polynomial with integer coefficients over Q and its group over Z_p, the finite field with p elements.

THEOREM 1. If F is a finite field, then there is a prime p such that for every $a \in F$ we have $\underbrace{a + \cdots + a}_{p \text{ times}} = 0$.

Proof. Let us write $j \cdot 1$ for $\underbrace{1 + \cdots + 1}_{j \text{ times}}$. Since F is finite, the elements $1 \cdot 1, 2 \cdot 1, \ldots$ cannot all be different. Suppose $j \cdot 1 = k \cdot 1$ and $j < k$. Then $0 = (k - j) \cdot 1$, so there is some integer n for which $n \cdot 1 = 0$.

Let r be the least positive integer such that $r \cdot 1 = 0$. If r factors into $r = st$ with $s, t \neq 1$, then we can write $r \cdot 1 = \underbrace{s \cdot 1 + \cdots + s \cdot 1}_{t \text{ times}}$, where $t < r$, Since $s \cdot 1 \in F$ it has an inverse, say b. Then $b(s \cdot 1 + \cdots + s \cdot 1) = \underbrace{1 + \cdots + 1}_{t \text{ times}} = t \cdot 1 = 0$, with $t < r$. Since r was assumed to be minimal, this is a contradiction, so r cannot factor and must be some prime p. It follows that for every a in F we have $p \cdot a = a + \cdots + a = a(1 + \cdots + 1) = a(p \cdot 1) = 0$. $\parallel$

DEFINITION. If p is a prime and for every $a \in F$ we have $p \cdot a = 0$, then p is called the *characteristic* of F. If $n \cdot 1 \neq 0$ for every natural number n, then we say that the *characteristic* of F is 0.

NOTATION. The characteristic of F is denoted by char F.

THEOREM 2. If F is finite, then there is a prime p and an integer n such that the number of elements in F is $|F| = p^n$.

Proof. For p let us take char F. This is a prime by Theorem 1. The elements 0, $1, \ldots, p - 1$ are necessarily all different, for if say $i \cdot 1 = j \cdot 1$, then $(j - i) \cdot 1 = 0$ and $j - i < p$, which contradicts Theorem 1. (We are assuming that $j > i$.)

Let F_0 be the field formed by $0, 1, \ldots, p - 1$ under addition and multiplication mod p. We can think of F as a vector space over F_0. Since F is finite, the dimension of F over F_0 must also be finite. Call it n and let $\alpha_1, \alpha_2, \ldots, \alpha_n$ be a basis for F over F_0. Then every $\alpha \in F$ can be expressed uniquely in the form

$$\alpha = a_1 \alpha_1 + \cdots + a_n \alpha_n,$$

where each $a_i \in F_0$. Since the coefficient of each a_i can therefore be chosen in p different ways, there are exactly p^n different expressions of this form, so we must have $|F| = p^n$. ‖

THEOREM 3. If char $F = p$, then for every $a, b \in F$ we have

$$(a + b)^{p^k} = a^{p^k} + b^{p^k}.$$

Proof. Expanding by the binomial theorem we have

$$(a + b)^{p^k} = a^{p^k} + \binom{p^k}{1}a^{p^k - 1}b + \cdots + \binom{p^k}{p^k - 1}ab^{p^k - 1} + b^{p^k} = a^{p^k} + b^{p^k},$$

because $p \,\Big|\, \binom{p^k}{j}$ for every j between 1 and $p^k - 1$. ‖

THEOREM 4. If char $F = p$, then every element a of $\dot{F}$ has a unique pth root $\sqrt[p]{a}$ in F.

Proof. We have for every $a, b \in F : (a - b)^p = a^p - b^p$, so

$$a - b = 0 \Leftrightarrow a^p - b^p = 0,$$

or, equivalently,

$$a = b \Leftrightarrow a^p = b^p.$$

Therefore, the elements $0, 1, a^p, b^p, c^p, \ldots$ of F are all different and so must be merely a rearrangement of $0, 1, a, b, c, \ldots$ of F and each element can be simply paired off with its pth root. ‖

THEOREM 5. If $|F| = p^n$, then F is the splitting field of the polynomial $x^{p^n} - x$.

Proof. If $|F| = p^n$, then $F^* = F - \{0\}$ has $p^n - 1$ elements. By the definition of a field, F^* is a group under multiplication and so the order of any nonzero element of F must divide the order of F^*, namely $p^n - 1$. In other words, $0 \neq a \in F \Rightarrow a^{p^n - 1} = 1$.

Therefore, every nonzero element of F is a root of $x^{p^n-1} - 1 = 0$, and every element of F is a root of $x^{p^n} - x = 0$. The polynomial $x^{p^n} - x$ therefore has p^n different roots and so it is separable.   ‖

THEOREM 6.   All fields containing p^n elements are isomorphic.

Proof.   They are all splitting fields of the separable polynomial $x^{p^n} - x$ and so are isomorphic by the first corollary on page 75 of Section 3.1.   ‖

NOTATION.   We use $GF(p^n)$, read as "the Galois field with p^n elements" to denote a splitting field of $x^{p^n} - x$ with characteristic p. By Theorem 6 all such fields are isomorphic. Also note that $GF(p) \cong Z_p$, the integers mod p.

THEOREM 7.   If $|F| = p^n$, then $F \rhd GF(p)$.

Proof.   By Theorem 2, char $F = p$, so $GF(p) < F$. By Theorem 5, F is the splitting field of a separable polynomial over $GF(p)$ and so it must be a normal extension.   ‖

THEOREM 8.   The nonzero elements of $GF(p^n)$ form a cyclic group under multiplication.

Proof.   If F^* is the group of nonzero elements of $GF(p^n)$, then $|F^*| = p^n - 1$, This group is Abelian and therefore the direct product of cyclic groups. Let $c_1, \ldots, c_k \in F^*$ be the generators of these cyclic groups. Every element $a \in F^*$ can then be expressed uniquely in the form $a = c_1^{i_1} \cdots c_k^{i_k}$ and we let $o(a)$ be its order and also let $m = \mathrm{lcm}(o(c_1), \ldots, o(c_k))$. The order of an element is a divisor of the order of the group, so $o(a)|(p^n - 1)$ and $m|(p^n - 1)$. Therefore,

$$a^m = c_1^{m i_1} \cdots c_k^{m i_k} = 1$$

for every a in F^*, which shows that the polynomial $x^m - 1$ must have $p^n - 1$ different roots in F^*. This implies that $p^n - 1|m$ and since also $m|(p^n - 1)$, we conclude that $m = p^n - 1$. Since $c_1, \ldots, c_k$ are the generators of F^*, we can have $(c_1 \cdots c_k)^h =$

4.11 FINITE FIELDS

$c_1^h \cdots c_k^h = 1$ only if $c_1^h = \cdots = c_k^h = 1$, so h must be a multiple of $o(c_i)$ for every i; that is, h must be a multiple of m. Therefore, $o(c_1 \cdots c_k) = m = p^n - 1$ and $c_1 \cdots c_k$ is therefore a generator of F^*, showing that F^* is cyclic. ∥

DEFINITION.   If $p(x) = \sum_{i=0}^{n} a_i x^i \in F[x]$, then its *formal derivative* is

$$p'(x) = \sum_{i=0}^{n} ia_i x^i.$$

THEOREM 9.   If $(x - a)^2 | p(x)$, then $(x - a) | p'(x)$.

Proof.   (1) Let

$$p(x) = (x - a)^2 q(x),$$

$$q(x) = \sum b_i x^i.$$

(2) Then on expanding and using the definition of the formal derivative we find that

$$p(x) = \sum b_i x^{i+2} - 2a \sum b_i x^{i+1} + a^2 \sum b_i x^i,$$

$$p'(x) = \sum (i + 2)b_i x^{i+1} - 2a \sum (i + 1)b_i x^i + a^2 \sum ib_i x^{i-1}$$

$$= x^2 \sum ib_i x^{i-1} + 2x \sum b_i x^i - 2ax \sum ib_i x^{i-1} - 2a \sum b_i x^i + a^2 \sum ib_i x^{i-1}$$

$$= (x - a)^2 q'(x) + 2(x - a)q(x).$$

(3) Therefore, $(x - a) | p'(x)$.   ∥

COROLLARY.   The polynomial $p(x)$ has a repeated root $\Leftrightarrow p(x)$ and $p'(x)$ have a common factor.

Proof.   Let E be the splitting field of $p(x)$.

$\Rightarrow$ If $p(x)$ has a repeated root, then it has a factor of the form $(x - a)^2$ in E. The rest follows from Theorem 9.

⇐ If $p(x)$ and $p'(x)$ have a common factor, then over E they have a common linear factor $x - a$. If, therefore, $p(x) = (x - a)r(x)$, a straightforward calculation similar to step (2) shows that $p'(x) = (x - a)r'(x) + r(x)$. By hypothesis, $(x - a)|p'(x)$, so we see that we must have $(x - a)|r(x)$, say $r(x) = (x - a)q(x)$. Substituting this into the expression for $p(x)$ gives $p(x) = (x - a)^2 q(x)$, so $p(x)$ has a repeated root. ‖

DEFINITION. A field F is called *perfect* if no irreducible polynomial over F has repeated roots. Otherwise it is called *imperfect*.

THEOREM 10. A field F is perfect ⇔

(1) char $F = 0$, *or*

(2) F is finite, *or*

(3) char $F = p$ and every element of F has a pth root that is also in F.

To prove this theorem we need the following lemma. The proof of Theorem 10 follows the proof of the lemma.

LEMMA. If $p(x)$ and $q(x)$ have a common factor of degree at least 1 when factored over the field K and $F < K$, then $p(x)$ and $q(x)$ already have a common factor when factored over F.

Proof of the Lemma. Suppose not. Then $p(x)$ and $q(x)$ would be relatively prime over F, so there would be some $a(x)$ and $b(x)$ in $F[x]$ such that $p(x) \cdot a(x) + q(x) \cdot b(x) = 1$. Since $F < K$, we also have $a(x), b(x) \in K[x]$, so this identity expresses the fact that $p(x)$ and $q(x)$ are relatively prime over K. This would contradict the hypothesis. ‖

Proof of Theorem 10. By definition,

F is imperfect ⇔ there is an irreducible polynomial $p(x) \in F[x]$ which has a repeated root.

4.11 FINITE FIELDS

By Theorem 9 and the preceding lemma, therefore,

F is imperfect $\Leftrightarrow$ there is an irreducible polynomial $p(x) \in F[x]$ such that $p(x)$ and
$p'(x)$ have a common factor $r(x)$ of degree at least 1.

However, $p(x)$ is irreducible and $p'(x)$ is of lower degree than $p(x)$, so $r(x)$ cannot be $p(x)$. Therefore, the only possibility is that $p'(x)$ is identically zero.

If $p(x) = \sum a_i x^i$, then

$$p'(x) = \sum i a_j x^j \text{ vanishes identically} \Leftrightarrow i a_i = 0 \text{ for every } i = 0, 1, 2, \ldots, n.$$

If char $F = 0$, then this implies that $a_i = 0$ for $i = 1, \ldots, n$ and $p(x) = a_0$. We conclude that a nontrivial irreducible polynomial cannot have double roots.

If char $F = p$, then $i a_i = 0$ for all i implies that $i = 0$ in F, or that $a_i = 0$ in F—that is, that $p|i$ or $a_i = 0$. Therefore, $p(x)$ will have repeated roots only if it is of the form

$$p(x) = a_0 + a_p x^p + a_{2p} x^{2p} + \cdots + a_{kp} x^{kp}.$$

If now every element of F has a pth root in F, then there is a b_i for every a_i such that $a_i = b_i^p$ and, therefore,

$$p(x) = b_o^p + b_p^p x^p + b_{2p}^p x^{2p} + \cdots + b_{kp}^p x^{kp}$$

$$= (b_0 + b_p x + b_{2p} x^2 + \cdots + b_{kp} x^k)^p,$$

so $p(x)$ is reducible. Therefore, in order that $p(x)$ be irreducible and also have repeated roots, there must be some coefficient that has no pth root in F.

Conversely, if some element a of F has no pth root in F and char $F = p$, then the polynomial $p(x) = x^p - a = (x - \sqrt[p]{a})^p$ clearly does have repeated roots (in some suitable extension K of F), although we can show that $x^p - a$ is irreducible over F: We prove this last fact by contradiction. Suppose $x^p - a$ did factor over F; then any factor $r(x)$ would have to be of the form

$$r(x) = (x - \sqrt[p]{a})^k,$$

for some k less than p, factored in the field K. Since the coefficients of $r(x)$ are in F, the constant term $a^{k/p}$ must be an element of F. Now k, being less than p, must be prime to p, and therefore there are integers r and s such that $rk + sp = 1$. Then $a^{rk/p}$ and $a^{sp/p} - a^s$ are all elements of F and so

$$a^{rk/p}a^{sp/p} = a^{(rk+sp)/p} = a^{1/p}$$

must also be in F, contrary to hypothesis. Therefore, $x^p - a$ is irreducible over F.

Combining both parts completes the proof of the theorem. $\|$

COROLLARY.   If char $F = p$, then

$p(x) \in F[x]$ has repeated roots $\Leftrightarrow p(x)$ is of the form $a_0 + a_1x^p + \cdots + a_kx^{kp}$.

Proof.   This was proved as part of the last theorem. $\|$

THEOREM 12.   If F is finite and char $F = p$, then $G(F/GF(p))$ is cyclic.

Proof.   We shall show that (1) the correspondence $\varphi : a \to a^p$ is an automorphism of F, and (2) every automorphism ψ of F is of the form $\psi = \varphi^{kp}$ for some k.

(1) Let $\varphi(a) = a^p$ for every α in F. Then

$$\varphi(a + b) = (a + b)^p = a^p + b^p = \varphi(a) + \varphi(b),$$

$$\varphi(ab) = (ab)^p = a^pb^p = \varphi(a)\varphi(b),$$

so φ is a homomorphism. By Theorem 4, φ is one to one and onto, so it is an iso-morphism.

(2) Let $|F| = p^n$, so $[F:GF(p)] = n$, and let c be a generator of the cyclic group of $F^* = F - \{0\}$ under multiplication. Then the n elements c^p, $c^{2p}, \ldots, c^{(n-1)p}$, $c^{np} = c$ are all different from each other and so the n automorphisms

$$\varphi : c \to c^p, \varphi^2 : c \to c^{2p}, \ldots, \varphi^n : c \to c^{np}$$

are all different. From the fundamental theorem we know that $n = [F:GF(p)] = |G(F/GF(p))|$, so these n automorphisms are the only ones possible. Thus $G(F/GF(p))$ must be the cyclic group generated by φ.   ‖

THEOREM 13.   The group $G(GF(p^{kn})/GF(p^n))$ is the cyclic group C_k.

Proof.   Exercise.

THEOREM 14.   If $|F| = p^n$ and $p(x) \in F[x]$ factors over F into k different irreducible factors,

$$p(x) = q_1(x) \cdot \ldots \cdot q_k(x),$$

where $\deg q_i(x) = n_i$ and K is its splitting field, then $G(K/F)$ is cyclic and is generated by a permutation containing k cycles with orders $n_1, \ldots, n_k$.

Proof.   If $K = F(\alpha_{1,1}, \ldots, \alpha_{k,n_k})$, where $\alpha_{i,1}, \ldots, \alpha_{i,n_i}$ are the roots of $q_i(x)$, then any automorphism of K will permute the roots of $q_i(x)$. As we already know that the group of $q_i(x)$ over F is cyclic, there must be a cycle of length n_i which generates this group. This holds for every $i = 1, \ldots, k$, so the group $G(K/F)$ is generated by the product of these disjoint cycles. (Notice that since the $q_i(x)$ are all different, no automorphism can send $a_{i,j}$ into any $\alpha_{h,l}$ with $i \neq h$.)   ‖

We next wish to show that the group of a polynomial $f(x)$ irreducible over Q and with integer coefficients taken mod p is a subgroup of the group of this same $f(x)$ over Q. Before we show this we state two reminders:

(1) If E is the splitting field of $f(x)$ over Q, then there are integers $m_1, \ldots, m_n$ such that

$$\gamma = m_1\alpha_1 + \cdots + m_n\alpha_n,$$

where $\alpha_1, \ldots, \alpha_n$ are the roots of $f(x)$ and $E = Q(\gamma)$. It is no restriction to think of $f(x)$ as having integer coefficients, because we can multiply through by the least common denominator of the coefficients without altering the roots.

(2) If $\gamma \in E$ is a root of the irreducible polynomial $p(x) \in Z[x]$, then the other roots of $p(x)$ are of the form

$$\gamma_\sigma = m_i\alpha_{\sigma(1)} + \cdots + m_n\alpha_{\sigma(n)},$$

where $\sigma \in G(E/Q)$ when this group is considered as a permutation group. In other words, $p(x)$ factors in E as

$$p(x) = \prod_{\sigma \in G(E/Q)} (x - \gamma_\sigma).$$

(See Section 4.8.) This is true because $\varphi(x) = \prod_{\sigma \in \mathfrak{S}_n} (x - \gamma_\sigma)$ has rational coefficients, since it is surely symmetric in the α's, and so any conjugate of γ must also be a root of this polynomial and thus be γ_σ for some σ. If $p(x)$ is the irreducible factor of $\varphi(x)$ which has γ as a root, then any automorphism σ of $G(E/Q)$ transforms γ into another root of $p(x)$, so all the conjugates of γ are roots of $p(x)$. Also any automorphism $\gamma \to \gamma_\sigma$, where γ_σ is a root of $p(x)$ is an automorphism of $G(E/Q)$, so if γ_σ is a root of $p(x)$, then $\sigma \in G(E/Q)$.

THEOREM 15. Let $f(x) \in Z[x]$ have roots $\alpha_1, \ldots, \alpha_n$ and E as splitting field and let $f^*(x) \in Z_p[x]$ be the image of $f(x)$ under the natural homomorphism φ_p of Z onto Z_p which sends every integer i into $i \pmod p$. If the roots of $f^*(x)$ are $\alpha_1^*, \ldots, \alpha_n^*$, then the splitting field of $f^*(x)$ is $Z_p(\alpha_1^*, \ldots, \alpha_n^*)$, and (as permutation groups on the roots)

$$G(E^*/Z_p) < G(E/Q).$$

Proof. (1) Let $\gamma = m_1\alpha_1 + \cdots + m_n\alpha_n$, with $m_1, \ldots, m_n$ in Z determined so that $E = Q(\gamma)$, and let γ^* be its image under the natural homomorphism φ_p, that is,

$$\gamma^* = m_1\alpha_1^* + \cdots + m_n\alpha_n^* \qquad \pmod p.$$

(2) Since γ generates all of E we must have $E^* = Z_p(\gamma^*)$.

(3) If $p(x)$ maps onto $p^*(x)$ and we can factor

$$p^*(x) = p_1^*(x) \cdots p_1^*(x) \qquad \pmod p$$

with each $p_i^*(x)$ irreducible, then γ^* must be a root of one of these, say of $p_1^*(x)$.

(4) Applying the argument of reminder (2) we see that the other roots of $p_1^*(x)$ are precisely the images γ_σ^* of γ_σ, where $\sigma \in G(E^*/Z_p)$.

(5) Therefore, $\sigma \in G(E^*/Z_p) \Rightarrow \sigma \in G(E/Q)$.   ‖

Note, however, that if, say, $\deg p_1^*(x) = 2$, we cannot immediately conclude that the cycle $(12) \in G(E/Q)$. For example, we may have $p_1^*(x) = p_2^*(x)$ and then σ is necessarily of the form $(12)(34)(\cdots)$, but we can be sure that $G(E/Q)$ will contain a permutation of order 2.

EXAMPLE.   Let $f(x) = x^5 - x - 1$ [vdW, p. 191].

(a) If we let $p = 2$, then $f^*(x) = (x^2 + x + 1)(x^3 + x^2 + 1)$. The group of $f^*(x)$ is cyclic, so the automorphisms of E^* will interchange the two roots α_1^* and α_2^* of $x^2 + x + 1$ and allow cyclic permutations of the roots $\alpha_3^*, \alpha_4^*, \alpha_5^*$ of $x^3 + x^2 + 1$. Therefore, $G(E^*/Z_p)$ contains a permutation of the form $(12)(345)$.

(b) Next we let $p = 3$. By trial and error we see that $f(x)$ is irreducible mod 3. Therefore, for $p = 3$ the group $G(E^*/Z_p)$ must be cyclic of order 5 and be generated by a cycle of the form (12345).

(c) Therefore, $G(E/Q)$ contains permutations of the form $(ij)(klm)$ and (12345), where i, j, k, l, m are the integers 1, 2, 3, 4, 5 in some order.

(d) We know that these together generate all of $\mathfrak{S}_5$; Surely $((ij)(klm))^3 = (ij)$ is in $G(E/Q)$ and so is $(12345)(ij)(12345)^{-1} = (\tau(i), \tau(j))$, where $\tau = (12345)^{-1}$. Similarly, the transpositions $(12345)^h(ij)(12345)^{-h} = (\tau^h(i), \tau^h(j))$ are also in $G(E/Q)$ and $i, \tau(i),$ $\tau^2(i), \tau^3(i), \tau^4(i)$ must be the integers 1, 2, 3, 4, 5 in some order. We have therefore generated five transpositions of the form $(ij), (jk), (kl), (lm), (mi)$ and all are in $G(E/Q)$.

(e) Therefore, all transpositions on $1, \ldots, 5$ are in $G(E/Q)$: for example, $(ik) = (ij)(jk)(ij)$. Since every permutation is a product of transpositions, this implies that $\mathfrak{S}_5 < G(E/Q)$.

(f) But also $G(E/Q) < \mathfrak{S}_5$, so $G(E/Q) = \mathfrak{S}_5$.

This is an excellent method for determining the group of a polynomial, provided it works. Unfortunately, it does not always yield sufficient information to determine the group completely.

EXAMPLE. The polynomial $f(x) = x^4 = 10x^2 + 1$ is irreducible over Q but it will factor into two (not necessarily irreducible) factors mod p for every p, so all we can tell about its group from this argument is that it must contain a permutation of the form (12)(34). By trial and error, for example, we find that

$x^4 - 10x^2 + 1 = ($_____$)($_____$)($_____$)($_____$)$ mod 2

$= ($_____$)($_____$)$ mod 3

$= ($_____$)($_____$)($_____$)($_____$)$ mod 23,

and so on. The roots of $f(x)$ are, in fact, $x = \pm\sqrt{2} \pm \sqrt{3}$, so the group of the polynomial is really the 4-group $\mathfrak{B}$. With this information it is easy to see why $f(x)$ must factor in this manner: If it were irreducible mod p for some p, then $G(E/Q)$ would contain a cycle of the form (1234), and if it had one linear and one cubic factor mod p, then $G(E/Q)$ would contain a cycle of the form (123). Since $\mathfrak{B}$ contains neither of these, $f(x)$ must always factor mod p, for every p.

EXERCISE. Show that if $p(x) \in Z[x]$ is of degree n and its group $G(E/Q)$ contains no cycle of the form $(12 \cdots n)$, then $p(x)$ is reducible mod p for every p.

4.12   MORE APPLICATIONS

In this section we shall use the fundamental theorem to prove some results that are not easily derived by other means. We leave it as an exercise to try to prove them in a straightforward manner and to realize the numerical complications that arise.

4.12 MORE APPLICATIONS

The first result and its proof are due to I. J. Richards of the University of Minnesota.

THEOREM. Let n be any integer, let $p_1 \ldots, p_k$ be distinct positive primes, and let $\sqrt[n]{a}$, for $a > 0$, denote the real positive nth root. Then the field $Q(\sqrt[n]{p_1}, \ldots, \sqrt[n]{p_k})$ is of degree n^k over Q.

Proof. Let

$\varepsilon =$ a primitive nth root of unity,

$R = Q(\varepsilon)$,

$\{\alpha_i\}$ be the set of n^k elements of the form

$$p_1^{r_1/n} \cdots p_k^{r_k/n} \qquad (0 \leqq r_j < n),$$

$\{\beta_i\}$, in the case where n is even, be the set of $(n/2)^k$ elements of the form

$$p_1^{r_1/n} \cdots p_k^{r_k/n} \qquad (0 \leqq r_j < n/2),$$

$S = Q(p_1^{1/2}, \ldots, p_k^{1/2})$,

$T = R(p_1^{1/2}, \ldots, p_k^{1/2})$.

The set $\{\alpha_i\}$ clearly spans $Q(\sqrt[n]{p_1}, \ldots, \sqrt[n]{p_k})$ over Q, and the theorem is equivalent to the statement that the α_i are linearly independent over Q.

Since the theorem for any multiple $n_i n$ clearly implies the theorem for n, there is no loss of generality in assuming that n is even. We shall prove that

(1) $[S:Q] = 2^k$.

(2) $[R(\sqrt[n]{p_1}, \ldots, \sqrt[n]{p_r}):T] = (n/2)^k$.

Since the set $\{\beta_i\}$ spans $Q(\sqrt[n]{p_1}, \ldots, \sqrt[n]{p_k})$ over S, (2) implies that

(3) $[Q(\sqrt[n]{p_1}, \ldots, \sqrt[n]{p_k}):S] = (n/2)^k$.

(1) and (3) imply our theorem. However, (2) is somewhat stronger, because it concerns R rather than Q. It is not true that $[T:R] = 2^k$ for all n. Thus, for instance,

if $n = 10$, then R contains $\sqrt{5}$. We shall show later in this section that for every prime p, $p \in Q(\varepsilon_p)$ or $p \in Q(i, \varepsilon_p)$, where ε_p is a primitive pth root of unity.

LEMMA 1. Let F be any field of characteristic zero. Suppose $a \in F$ and that the polynomial $x^n - a$ factors over F. Then there is an $m|n$, $m > 1$, an nth (not necessarily mth) root of unity ε and one value of $\sqrt[m]{a}$ such that $\varepsilon\sqrt[m]{a} \in F$.

Proof. Write $x^n - a = (x - \sqrt[n]{a})(x - \varepsilon\sqrt[n]{a}) \cdots (x - \varepsilon^{n-1}\sqrt[n]{a})$. The constant term in any factor of $x^n - a$ has the form $\pm \varepsilon^c a^{r/n} \in F$ ($0 < r < n$). Let $s = \gcd(r, n)$ and $m = n/s$. Take integers M, N so that $Mn + Nr = s$. Then $b = a^M(\varepsilon^c a^{r/n})^N \in F$, and $b/\varepsilon^{cN} = a^{s/n} = a^{i/m}$. ‖

LEMMA 2. For any n, the Galois groups of R over Q and of T over Q are Abelian.

Proof. For R we already know that $G(R/Q) \cong$ (the multiplicative group of integers prime to n), Now by direct examination of the possible automorphisms, it is clear that $G(T/Q)$ has the form $G(R/Q) \oplus Z_2 \oplus \cdots \oplus Z_2$. (The group Z_2 occurs some number $j \leq k$ times.) ‖

LEMMA 3. When m is prime and $R' = Q(\varepsilon_m)$, then $[R' : Q] = m - 1$.

Proof. Apply the Eisenstein irreducibility theorem and the substitution $x = y + 1$ to the polynomial $(x^m - 1)/(x - 1)$. ‖

LEMMA 4. (a) If m is prime and $a \in Q$ has no rational mth root, then $x^m - a$ is irreducible over $R' = Q(\varepsilon_m)$.
  (b) For $m = 4$: if $a > 0$ and $a \in Q$ has no rational square root, then $x^4 - a$ is irreducible over $R' = Q(i)$.

Proof. For (a), combine Lemmas 1 and 3. For (b), use Lemma 1 and the fact that all the real numbers in $Q(i)$ are rational. ‖

(A different proof of Lemma 4 will be found in [vdW, p. 171].)

4.12   MORE APPLICATIONS

LEMMA 5.   (a) If $m > 2$ is prime and $a \in Q$ is as in Lemma 4(a), then the Galois group of $R'(\sqrt[m]{a})$ over Q is not Abelian.

(b) For $m = 4$, $a \in Q$ as in Lemma 4(b), the Galois group of $R'(\sqrt[m]{a})$ over Q is not Abelian.

Proof.   For (a): Since by Lemmas 3 and 4(a), $[R'(\sqrt[m]{a}):Q] = m(m - 1)$, all the automorphisms generated by the maps

$$\phi_r : \varepsilon \to \varepsilon^r, \sqrt[m]{a} \text{ held fixed } (0 < r < m),$$

$$\psi_s : \sqrt[m]{a} \to \varepsilon^s, \sqrt[m]{a} \text{ held fixed } (0 \leqq s < m),$$

actually occur. But $\phi_2\psi_1 \neq \psi_1\phi_2$.

The proof of (b) is similar.   ‖

LEMMA 6.   Here n is arbitrary, R and T are as above. Suppose that $m > 2$, $a > 0$, $a \in Q$, and a has no qth root in Q for any $q \mid m, q > 1$. Then $\sqrt[m]{a} \notin T = R(p^{1/2}, \ldots, p_k^{1/2})$.

Proof.   It suffices to consider only the two cases $m = \text{prime} > 2$, and $m = 4$. Thus, since T is a normal extension of Q, Lemma 6 follows from Lemmas 2 and 5.   ‖

We have already seen that (2) implies (3). To prove (2), we take a fixed $M \geq k$ and let $T^* = R(p_1^{1/2}, \ldots, p_k^{1/2})$. T^* will now remain fixed and we set $F_k = T^*(p_1^{1/n}, \ldots, p_k^{1/n})$, where $k \leq M$.

We shall prove by induction that $[F_k : T^*] = (n/2)k$.

For $k = 1$, this follows from Lemmas 1 and 6. (The polynomial $x^{n/2} - p_1^{1/2}$ is irreducible over T^*.)

Assume now that $[F_k : T^*] = (n/2)^k$. We wish to show that $x^{n/2} - p_{k+1}^{1/2}$ is irreducible over F_k.

Suppose not. Then by Lemma 1 there is an integer $m \mid (n/2)$, $m > 1$, such that $p_{k+1}^{1/2m} \in F_k$.

By our induction hypothesis, the set $\{\beta_i\}$ defined above forms a basis for F_k over T^*. Thus

$$p_{k+1}^{1/2m} = c_1\beta_1 + \cdots + c_N\beta_N, \text{ where } c_i \in T^* \text{ and } N = (n/2)^k.$$

There are now two cases:

Case 1. Exactly one of the $c_i \neq 0$. This contradicts Lemma 6.

Case 2. We have c_i and $c_j \neq 0$, $i \neq j$. Since $(\beta_i/\beta_j) \notin T^*$, there exists an automorphism ϕ of F_k over T^* such that $\phi(\beta_i/\beta_j) \neq \beta_i/\beta_j$. But since all the β's and also $p_{k+1}^{1/2m}$ are nth roots of integers,

$$\phi(\beta_i) = \varepsilon^r \beta_i,$$

$$\phi(\beta_j) = \varepsilon^s \beta_j, \qquad \text{where } \varepsilon^s \neq \varepsilon^r,$$

$$\phi(\beta_h) = \varepsilon^{c(h)} \beta_h, \qquad \text{for } h \neq i, j,$$

$$\phi(p_{k+1}^{1/2m}) = \varepsilon^t p_{k+1}^{1/2m},$$

for some r, s, $c(h)$, t. Then, since $\varepsilon \in T^*$, applying ϕ to the linear equation for $p_{k+1}^{1/2m}$ contradicts the linear independence of $\{\beta_i\}$ over T^*. This proves (2).

To prove (1), we can repeat the above proof of (2) replacing T^* by Q. As an alternative, it is not difficult to give a purely computational induction proof for (1). ‖

This result shows that $\sqrt[n]{p_1}$ is not expressible rationally in terms of other nth roots, so, for example, $\sqrt[3]{2}$ is not a rational combination of the cube roots of other primes. There are many other similar theorems, as, for example, the following:

THEOREM. $\sqrt[3]{2}$ is not a rational combination of square roots of rational numbers.

Proof. Take $n = 6$ in Richard's theorem. ‖

Similarly, $\sqrt[3]{2}$ is not a rational combination of fifth roots, and so on.

We now return to the cyclotomic field Q_ε obtained from the rationals by adjoining

4.12 MORE APPLICATIONS

all nth roots of unity ε_n for every n. We already know that the group $G(Q(\varepsilon_n)/Q)$ is Abelian, so the next theorem could have been used in Richard's proof to show that $G(Q(p_1^{1/2}, \ldots, p_n^{1/2})/Q)$ is Abelian.

THEOREM. $Q(\sqrt{2}, \sqrt{3}, \sqrt{5}, \ldots, \sqrt{n}, \ldots) < Q_\varepsilon$.

Proof.   It is clearly sufficient to show that $\sqrt{p} \in Q_\varepsilon$ for every prime p.

(1) Let p be a prime, $p > 2$, and let

$$\Phi(x) = x^{p-1} + \cdots + x + 1$$

be the pth cyclotomic polynomial. The discriminant Δ of $\phi(x)$ is

$$\Delta = \prod_{i<j} (\varepsilon^i - \varepsilon^j)^2, \qquad \text{where } i, j = 1, \ldots, p - 1$$

and ε is a primitive pth root of unity.

(2) Since $\varepsilon, \varepsilon^2, \ldots$ are the roots of $\Phi(x)$, we have

$$\Phi(x) = (x - \varepsilon)(x - \varepsilon^2) \cdots (x - \varepsilon^i) \cdots (x - \varepsilon^{p-i}),$$

and on differentiation with respect to x we get

$$\Phi'(x) = \sum_{i=1}^{p-1} (x - \varepsilon) \cdots (x - \varepsilon^{i-1})(x - \varepsilon^{i+1}) \cdots (x - \varepsilon^{p-1}),$$

$$\Phi'(\varepsilon^i) = (\varepsilon^i - \varepsilon) \cdots (\varepsilon^i - \varepsilon^{i-1})(\varepsilon^i - \varepsilon^{i+1}) \cdots (\varepsilon^i - \varepsilon^{p-1}).$$

(3) Multiplying all the $\Phi'(\varepsilon^i)$ together we get

$$\prod_i \Phi'(\varepsilon^i) = [(\varepsilon - \varepsilon^2)(\varepsilon - \varepsilon^3) \cdots (\varepsilon - \varepsilon^{p-1})]$$

$$\cdot [(\varepsilon^2 - \varepsilon)(\varepsilon^2 - \varepsilon^3) \cdots (\varepsilon^2 - \varepsilon^{p-1})]$$

$$\cdots$$

$$\cdot [(\varepsilon^{p-1} - \varepsilon)(\varepsilon^{p-1} - \varepsilon^2) \cdots (\varepsilon^{p-1} - \varepsilon^{p-2})]$$

$$= (-1)^{p(p-1)/2} \Delta.$$

(4) But we also have

$$x^p - 1 = (x - 1)\Phi(x),$$

and on differentiating this we get

$$px^{p-1} = \Phi(x) + (x - 1)\Phi'(x).$$

(5) This gives

$$p\varepsilon^{i(p-1)} = (\varepsilon^i - 1)\Phi'(\varepsilon^i),$$

since $\Phi(\varepsilon^i) = 0$.

(6) Now multiply all these together:

$$p^{p-1}\varepsilon^{(p-1)}\varepsilon^{2(p-1)} \cdots \varepsilon^{(p-1)(p-1)} = \prod_i (\varepsilon^i - 1) \prod_i \Phi'(\varepsilon^i),$$

and using the fact that $\varepsilon^{(p-1)} \cdots \varepsilon^{(p-1)(p-1)} = (\varepsilon, \varepsilon^2, \ldots, \varepsilon^{p-1})^{p-1} = 1^{p-1} = 1$, while $\prod_i (\varepsilon^i - 1) = (-1)^{p-1}\Phi(1)$, we get

$$p^{p-1} = (-1)^{p-1}p \cdot \prod_i \Phi'(\varepsilon^i),$$

$$\prod_i (\varepsilon^i - 1) = \prod_i (1 - \varepsilon^i) = \Phi(1) = p,$$

because $p - 1$ is even.

(7) Combining this with (3) gives

$$p^{p-2} = (-1)^{p(p-1)/2}\Delta = (-1)^{p-1/2}\Delta,$$

since $(-1)^{p(p-1)/2} = ((-1)^p)^{p-1/2} = (-1)^{p-1/2}$.

(8) Therefore,

$$\Delta = (-1)^{p-1/2}p^{p-2}$$

$$= \begin{cases} p^{p-2}, & \text{if } \dfrac{p-1}{2} \text{ is even}, \\[2em] -p^{p-2}, & \text{if } \dfrac{p-1}{2} \text{ is odd}. \end{cases}$$

4.12 MORE APPLICATIONS

(9) However,

$$\frac{p-1}{2} \text{ is even} \Leftrightarrow p \equiv 1 \ (\text{mod } 4),$$

$$\frac{p-1}{2} \text{ is odd } \Leftrightarrow p \equiv 3 \ (\text{mod } 4).$$

(10) Combining this with (8) we have

$$\Delta = \begin{cases} p \cdot p^{p-3}, & \text{if } p \equiv 1 \ (\text{mod } 3), \\ \\ -p \cdot p^{p-3}, & \text{if } p \equiv 3 \ (\text{mod } 4), \end{cases}$$

$$\sqrt{\Delta} = \begin{cases} p^{(p-3)/2} \cdot \sqrt{p}, & \text{if } p \equiv 1 \ (\text{mod } 4), \\ \\ i p^{(p-3)/2} \cdot \sqrt{p}, & \text{if } p \equiv 3 \ (\text{mod } 4). \end{cases}$$

(11) By the definition of Δ it is the square of an element of the field $Q(\varepsilon_p)$, so $\sqrt{\Delta} \in Q(\varepsilon_p)$. Note also that $(p - 3)$ is even, making $p^{(p-3)/2}$ an integer.

As a result we have

$$\sqrt{p} = \begin{cases} \sqrt{\Delta}/p^{(p-3)/2} \in Q(\varepsilon_p), & \text{if } p \equiv 1 \ (\text{mod } 4), \\ \\ -i\sqrt{\Delta}/p^{(p-3)/2} \in Q(i, \varepsilon_p), & \text{if } p \equiv 3 \ (\text{mod } 4), \end{cases}$$

and $Q(i, \varepsilon_p) = Q(\varepsilon_{2p})$.

(12) We must still show that $\sqrt{2} \in Q$. Evidently,

$$\varepsilon^8 = e^{2\pi i/8} = \cos(\pi/4) + i \sin(\pi/4) = \frac{\sqrt{2}}{2}(1 + i),$$

so

$$\sqrt{2} = \frac{2\varepsilon_8}{1 + i} = \frac{2\varepsilon_8(1 - i)}{(2 + i)(1 - i)} = \varepsilon_8(1 - i) \in Q.$$

(13) Summarizing, we have

$$\sqrt{2} \in Q(\varepsilon_8),$$

$$\sqrt{p} \in Q(\varepsilon_p), \qquad \text{if } p \equiv 1 \pmod 4,$$

$$\sqrt{p} \,\varepsilon\, Q(\varepsilon_{2p}), \qquad \text{if } p \equiv 3 \pmod 4.$$

(14) In any case, therefore, we have shown that

$$Q(\sqrt{2}, \sqrt{3}, \ldots) < Q_\varepsilon. \qquad \|$$

COROLLARY. If $p > 2$ and $a \in Q$ has no rational pth root, then $a^{1/p} \notin Q_\varepsilon$, but $a^{1/2} \in Q_\varepsilon$.

Proof. Note that $a^{i/p} \in Q_\varepsilon$ would imply that $G(Q(\varepsilon_p, a^{1/p}))$ be Abelian. $\qquad \|$

As we saw earlier, not all nth roots of unity are constructible by ruler and compass, but some of them certainly are. One might ask: Is every constructible complex number a rational function of some appropriate nth roots of unity? Or, equivalently, is the field K of constructible complex numbers a subfield of the cyclotomic field Q_ε? We know that K is closed under the square-root operation and that $\sqrt{n} \in Q$ for every $n \in Z$, but is Q_ε closed under $\sqrt{\ }$? For example, can we show that $\sqrt{1 + \omega} \in Q$? We do have $1 + \omega + \omega^2 = 0$, so that $1 + \omega = -\omega^2$ and $\sqrt{1 + \omega} = i\omega$, which is in Q_ε, but this is a long way from proving that Q_ε is closed under $\sqrt{\ }$. In fact, Q_ε is not closed under $\sqrt{\ }$ (see below). This is easily shown by methods similar to the preceding ones.

We have

THEOREM. If Q_ε is the cyclotomic field and K is the field of all those algebraic numbers which are constructible by ruler and compass, then

(1) $Q_\varepsilon \not< K,$

(2) $K \not< Q_\varepsilon,$

(3) $\mathscr{C} \lneqq Q_\varepsilon \cap K,$

where $\mathscr{C}$ is the field of complex rationals, $\mathscr{C} = Q(i)$ (see Figure 4.5).

4.12   MORE APPLICATIONS

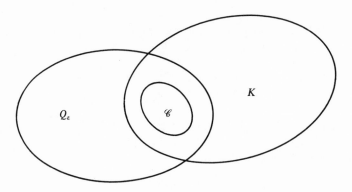

FIGURE 4.5

Proof. (1) This part is only a restatement of our previous result that not all regular *n*-gons are constructible.

(2) Note that $\sqrt[4]{2}$ is constructible. If, however, $\sqrt[4]{2} \in Q_\varepsilon$, then $Q(i, \sqrt[4]{2}) \lhd Q(\varepsilon_n)$ for some *n*, and

$$G(Q(i, \sqrt[4]{2})/Q) = G(Q(\varepsilon_n)/Q \,/\, G(Q(\varepsilon_n)/Q(i, \sqrt[4]{2})).$$

Therefore, the non-Abelian group $G(Q(i, \sqrt[4]{2})/Q)$ would be a homomorphic image of the Abelian group $G(Q(\varepsilon_n)/Q)$, and this is impossible.

(3) This is trivial, because $\omega \notin \mathscr{C}$ but $\omega \in Q_\varepsilon \cap K$.   ‖

The last result we shall prove is a beautiful theorem which characterizes all Abelian extensions of *Q*. This was first stated by Kronecker and proved by Weber:

THEOREM.   If $E \rhd Q$ and $G(E/Q)$ is Abelian, then $E < Q(\varepsilon_m)$ for some integer *m*. Or, every finite Abelian extension of *Q* is a subfield of the cyclotomic field Q_ε.

Proof. (1) If $G(E/Q)$ is Abelian, then it must surely be solvable, so we know that there is a sequence of fields

$$Q = F_0 < F_1 \cdots < F_{r-1} < F_r < \cdots < F_n = E,$$

where $[F_r : F_{r-1}] = p_r$, a prime for $r = 1, \ldots, n$.

(2) We proceed by induction. Clearly $Q < Q_\varepsilon$. Suppose $F_{r-1} < Q_\varepsilon$; we must show that $F_r < Q_\varepsilon$.

(3) Since $E \rhd Q$ and $G(E/Q)$ is Abelian, each subgroup of $G(E/Q)$ is a normal subgroup and its fixed field is a normal extension of Q; therefore, $F_r \rhd Q$, for every $r = 0, 1, \ldots, n$.

(4) Let $F_r = F_{r-i}(\gamma)$, where $\gamma^{p_r} = \alpha \in F_{r-1}$.

(5) Let $\alpha = \alpha_1$ and let $\alpha_1, \alpha_2, \ldots, \alpha_k$ be the conjugates of α over Q; that is, for every i between i and k there is an automorphism $\varphi \in G(F_{r-1}/Q)$ such that $\varphi(\alpha_1) = \alpha_i$.

(6) For each i, let $\gamma_i, \varepsilon\gamma_i, \ldots, \varepsilon^{p-1}\gamma_i$ be the pth roots of α_i, where $p = p_r$ and $\varepsilon = \varepsilon_p = \sqrt[p]{1}$, and let γ_1 be γ.

(7) Since $F_r = F_{r-1}(\gamma_1)$, $\gamma_1^p \in F_{r-1}$, and $[F_r : F_{r-1}] = p \neq 1$, there must be an automorphism $\varphi_1 \in G(F_r/F_{r-1})$ such that φ_1 sends γ_1 into $\varepsilon\gamma_1 \neq \gamma_1$. Moreover, $F_r = F_{r-1}(\gamma_1) = F_{r-1}(\varepsilon\gamma_1)$, $F_r \rhd Q$, $F_{r-1}(\varepsilon\gamma_1) \rhd Q$, and so each of these extensions must contain all the conjugates of γ_1.

(8) There must also be an automorphism $\psi \in G(F_{r-1}/Q)$ for which $\psi(\alpha_1) = \alpha_2$ and $\alpha_2 \neq \alpha_1$, unless α_1 is fixed under all $\psi \in G(F_{r-1}/Q)$. In the latter case $\alpha_1 \in Q$. We shall discuss this case later in step (22).

(9) Suppose, therefore, that $\alpha_1 \neq \alpha_2$. Since $F_r \rhd Q$, ψ can be extended to an automorphism $\varphi_2 \in G(F_r/Q)$ and we have

$$\varphi_2 : \begin{cases} \alpha_1 \to \alpha_2 \neq \alpha_1, & \\ \gamma_1 \to \gamma_2 \neq \gamma_1, & \text{since } \gamma_1^p = \alpha_1 \neq \alpha_2 = \gamma_2^p, \\ \varepsilon \to \varepsilon^j, & \text{for some } j, 1 \leq j \leq p. \end{cases}$$

(10) We then have

$$\varphi_1\varphi_2(\gamma_1) = \varphi_1(\gamma_2),$$

$$\varphi_2\varphi_1(\gamma_1) = \varphi_2(\varepsilon\gamma_1) = \varphi_2(\varepsilon)\varphi_2(\gamma_1) = \varepsilon^j\gamma_2.$$

4.12 MORE APPLICATIONS

(11) Since $G(F_r/Q)$ is Abelian,

$$\varphi_1\varphi_2 = \varphi_2\varphi_1,$$

so that

$$\varphi_1(\gamma_2) = \varepsilon^j\gamma_2, \qquad \text{for some } j, 1 \le j \le p.$$

(12) Again since $F_r = F_{r-1}(\gamma_1) \rhd F_{r-1}$ and γ_2 is conjugate to γ_1, we know that $\gamma_2 \in F_{r-1}(\gamma_1)$, and so

$$\gamma_2 = c_0 + c_1\gamma_1 + \cdots + c_j\gamma_1^j + \cdots + c_{p-1}\gamma_1^{p-1},$$

with $c_i \in F_{r-1}$ uniquely determined.

(13) Apply φ_1 and remember that $\varphi_1 \restriction F_{r-1} = 1$, so $\varphi_1(c_i) = c_i$, for every i. Then

$$\varphi_1(\gamma_2) = c_0 + c_1\varepsilon\gamma_1 + \cdots + c_j\varepsilon^j\gamma_1^j + \cdots + c_{p-1}\varepsilon^{p-1}\gamma_1^{p-1},$$

and, by (11) and (12),

$$\varphi_1(\gamma_2) = \varepsilon^j\gamma_2 = c_0\varepsilon^j + \cdots + c_j\varepsilon^j\gamma_1^j + \cdots + c_{p-1}\varepsilon^j\gamma_1^{p-1}.$$

(14) Therefore, $c_0 = \cdots = c_{j-1} = c_{j+1} = \cdots = c_{p-1} = 0$ and $c_j = 1$; that is, $\gamma_2 = \gamma_1^j$, where $j \ne 1$, since $\gamma_1 \ne \gamma_2$.

(15) Notice that since γ_1 and γ_2 are conjugates, we have $|\gamma_1| = |\gamma_2|$, and by (14), also $|\gamma_2| = |\gamma_1^j| = |\gamma_1|^j$.

(16) Moreover, $j \ne 1$, so combining (14) and (15) gives

$$|\gamma_1| = |\gamma_1|^j, j \ne 1, \text{ and } |\gamma_1| \ne 0, \text{ since } F_r \ne F_{r-1}.$$

(17) Therefore $|\gamma_1| = 1$.

(18) Similarly, we have $\gamma_1 = \gamma_1^{j_i}$, for $2 \le i \le k$.

(19) Multiplying all the α's we get $\alpha_1 \cdot \ldots \cdot \alpha_k = \gamma_1^p \cdot \ldots \cdot \gamma_k^p = \gamma_1^N$, for some N, and

$$A = \alpha_1 \cdot \ldots \cdot \alpha_k \in Q.$$

(20) By (17) we have $|A| = |\gamma_1^N| = 1$, so $A = \pm 1$ and γ_1 is a root of unity.

(21) Since by our induction hypothesis $F_{r-1} < Q(\varepsilon_s)$ for some s and $F_r = F_{r-1}(\varepsilon_1)$, so we must have $F_r < Q(\varepsilon_m)$ for some m.

(22) This proves the theorem if $\alpha_1 \neq \alpha_2$, and we now return to the case $\alpha = \alpha_1 = \alpha_2 \in Q$. There are two subcases:

(1) $p = 2$,

(2) $p > 2$.

(23) If $p = 2$, then $\gamma^2 = \alpha \in Q$ and in this case we already know from preceding results that $\gamma \in Q(\varepsilon_s)$ for some integer s, so $F_{r-1}(\gamma) = F_r < Q(\varepsilon_N)$ for some N.

(24) If $p < 2$, then $F_r = F_{r-1}(\gamma)$ also contains $\varepsilon, \varepsilon^2, \ldots, \varepsilon^{p-1}$, since it is normal over Q, and $\varepsilon \neq \varepsilon^2$.

(25) As in the proof of Richard's theorem, we can then show that there are two automorphisms φ_1 and φ_2 such that

$$\varphi_1 : \begin{cases} \varepsilon \to \varepsilon \\ \\ \gamma \to \varepsilon\gamma \end{cases} \quad \text{and} \quad \varphi_2 : \begin{cases} \varepsilon \to \varepsilon^2 \\ \\ \gamma \to \gamma \end{cases}$$

and $\varphi_1\varphi_2 \neq \varphi_2\varphi_1$.

(26) For by hypothesis $[F_r : F_{r-1}] = p$ and so $x^p - \alpha$ is irreducible over F_{r-1} and the splitting field of $x^p - \alpha$ contains all $\varepsilon^j\gamma$.

(27) To check that $\varphi_1\varphi_2 \neq \varphi_2\varphi_1$, we calculate

$$\varphi_1\varphi_2(\gamma) = \varphi_1(\gamma) = \varepsilon\gamma,$$

$$\varphi_2\varphi_1(\gamma) = \varphi_2(\varepsilon\gamma) = \varepsilon^2\gamma,$$

and $\varepsilon \neq \varepsilon^2$.

(28) Therefore $G(F_r/Q)$ would not be Abelian, which would contradict our basic hypothesis that $G(E/Q)$ is Abelian, so this case cannot occur.

(29) Combining steps (21), (23), and (28) proves the theorem. $\parallel$

4.12   MORE APPLICATIONS

Since $\varepsilon_m = e^{2\pi i/m}$, it is a special value of the function $e^{2\pi iz}$, it is natural to ask for generalizations of this function which play a similar role for other algebraic number fields. This is Hilbert's twelfth problem. A special case was first proposed and solved by Kronecker, and the general case later completed by Weber, Takagi, and Hasse. It leads to a study of elliptic and of elliptic modular functions and to class-field theory. [See G. Shimura, *Automorphic Functions and Number Theory*, Springer Lecture Notes, Vol. 54, Springer-Verlag, Berlin (1968), or J. W. S. Cassels and A. Fröhlich, eds., *Algebraic Number Theory*, Thompson Book Co., Washington, D.C., 1967.] But that is another story.

BIBLIOGRAPHY

[Ad] Adamson, I. T., *Introduction to Field Theory*, University Mathematical Texts, Wiley, New York, 1964.

[A] Artin, E., *Galois Theory*, Notre Dame Mathematical Lectures, Notre Dame, Ind., 1942.

[A2] Artin, E., *Modern Higher Algebra: Galois Theory* (notes by A. A. Blank), NYU–Courant Institute of Mathematical Sciences, New York, 1947.

[C–R] Courant, R., and Robbins, H., *What Is Mathematics?* Oxford University Press, New York, 1941.

[D] Dean, R. A., *Elements of Abstract Algebra*, Wiley, New York, 1966.

[Di] Dickson, L. E., *Algebraic Theories*, Dover, New York, 1959 (orig. ed. 1926).

[H] Herstein, I., *Topics in Algebra*, Blaisdell, Waltham, Mass., 1964.

[J] Jacobson, N., *Lectures in Abstract Algebra*, Vol. III, Van Nostrand Reinhold, New York 1964.

[MacD] MacDuffee, C. C., *An Introduction to Abstract Algebra*, Wiley, New York, 1940.

[vdW] Van der Waerden, B. L., *Modern Algebra*, Vol. I, Ungar, New York, 1949 (orig. ed. 1931).

[W] Weisner, L., *Theory of Equations*, Macmillan, New York, 1938.

INDEX

INDEX